The MacGyver Secret

Connect to Your Inner MacGyver And Solve Anything

By Series Creator
Lee David Zlotoff

With
Colleen Seifert, Ph.D.

Copyright © 2016 MacGyver Solutions LLC

All rights reserved.
Published in the United States by MacGyver Solutions LLC, Delaware

ISBN: 978-0-692-76144-1
Printed in the United States of America

Distributed in the United States and Internationally
by Ingram Spark, 1 Ingram Blvd, La Vergne, TN 37086

First Edition

The MacGyver Foundation: A portion of the sales from this book and every MacGyver project, goes to support *The MacGyver Foundation*. www.macgyverglobal/foundation

ALL INQUIRIES regarding this book and the authors should be directed to: www.MacGyverSecret.com

Cover Design: *Shawn Patrick*
Back Cover Photograph: *Martin Christopher Photography*

To Dayna,
For making all the pieces
fit... in every way

Preface

A WORD ABOUT GENDER

The MacGyver character on TV is a man. But, whether you're a man or a woman, your 'inner MacGyver' referred to throughout this book *is neither specifically male or female.* Hence, I have tried as much as possible to refer to your inner MacGyver throughout the book as 'it' rather than 'he'.

Because everything proposed in this book applies just as much to women as it does to men. Or however you wish to identify your gender.

So, though most presume the term 'MacGyver' is inherently masculine, please take no inference—or offense, because I intend neither.

Bottom line, *if you're human, you have an inner MacGyver.* So, I thank you in advance for your patience and understanding with this minor, cultural dilemma.

Ideally, once I manage to create a female MacGyver in the world of media, this will cease to be an issue entirely. I'm working on it.

Contents

Preface 4

PART 1 — Introducing The MacGyver Secret 7
1. What is the MacGyver Secret? 9
2. How I Discovered The MacGyver Secret 11
3. How I Put My Inner MacGyver to Work 15
4. How The MacGyver Secret Can Work for You 19

PART 2 — The Three Steps of The MacGyver Secret 25

STEP ONE
5. Getting Started 27
6. How to Ask the Right Questions - Or Asking Questions Rightly 39

STEP TWO
7. Letting Your Inner MacGyver Get to Work 51
8. What is My Inner MacGyver, Really? 65

STEP THREE
9. Asking Your Inner MacGyver for the Answers 73
10. How to Best Evaluate the Answers From Your Inner MacGyver 79

PART 3 — The MacGyver Secret In Action 87
11. How to Make the MacGyver Secret a Habit 89
12. Using Sleep to Let Your Inner MacGyver Do Its Thing ... 97
13. Asking Your Inner MacGyver Tough Questions 103
14. Using The MacGyver Secret in the Workplace 109
15. Using The MacGyver Secret With a Team 113
16. Feeding Your Inner MacGyver 119
17. Life With Your Inner MacGyver 125
18. What's Next? 133

Acknowledgements 135
Cheat Sheet 137
References: *End Notes & Bibliography* 138
Index: *Science & Aha Stories* 145
About the Authors 146
Online Training Course, Presentations & Workshops 147

"The intuitive mind is a sacred gift and the rational mind is a faithful servant. We have created a society that honors the servant and has forgotten the gift."

— *Albert Einstein*

PART 1

Introducing The MacGyver Secret

The MacGyver Secret

Chapter 1
What is The MacGyver Secret?

You *know* MacGyver. That icon of "calm in a crisis." The very embodiment of resourcefulness and ingenuity who can overcome any obstacle by using whatever lay at hand. The absolute epitome of coming up with a brilliant solution when seemingly all looks hopeless. And doing it all with a smile. Right?

Of course, that MacGyver is just a fiction. Sure, you've probably done a 'MacGyver' or two, using a bit of tape or a paperclip to fish your keys out of a crack or keep that annoyingly loose car door from flying open on the highway.

But to consistently come up with clever solutions when your back is against the wall and everything seems to be falling apart? Again and again? How many of us really have that kind of genius? How many of us can be like MacGyver? No, that's got to be just a fantasy cooked up by some smartass TV writer, no?

Or does it?

What if it turned out that *anyone* could come up with a truly creative solution when faced with a tough problem? What if that was something even YOU could do? Just like MacGyver. Whenever you needed it. Whenever you wanted. Anywhere. Again and again. Whether you're an entrepreneur, or an engineer, or a designer, or a writer, or a secretary or *whatever* you might be.

What if it turned out that everyone—including YOU—had a "MacGyver" hidden deep inside? Just waiting to be handed a problem so it could whip up an amazing solution. For virtually any problem. Be it technical, creative, professional, personal, or *anything!* That would be another fantasy, you'd think.

Well, it's not. And that is THE MACGYVER SECRET.

Because the fact is, **each one of us has a MacGyver within us** ready to jump in and lend a hand whenever we need it.

And, in this book, I'll not only tell you how I discovered my inner MacGyver and put it to work for me, but how others have used this secret to achieve their breakthroughs. Most importantly though, I'll show you exactly how, with a few simple steps, you can *unlock that MacGyver inside you.* To tackle anything.

Along with the science to *prove* that it works.

With nothing more than a pen and a piece of paper.

Sounds like a challenge worthy of MacGyver, huh?

But, really, the only question for you right now is this: ***Are you ready to find your inner MacGyver?***

Chapter 2

My Story: How I Discovered The MacGyver Secret

It was sometime in the late 70's, after having spent a fruitful but frustrating year writing dialog for a soap opera in New York. I convinced my then-pregnant wife that our destiny— and fortune—lay waiting for us in the entertainment capital of Los Angeles. No doubt I was brimming with the confidence of someone in their mid-twenties, and had squirreled away enough soap writing funds to effect such a brash plan. We knew almost no one in LA— and not a single soul who was actually in show business. But move we did.

And, after a few years of trying to break into writing for TV—and by then the father of two children—the golden door finally swung wide and I landed a job on the writing staff for a prime-time TV series. That was the good news.

The bad news: Despite my soap opera training, I suddenly found myself having to crank out not only dialog but stories, outlines and scripts for episodes at a feverish pace to keep up with the relentless demands of TV production. This was a near constant stream of creative material that I had to produce in very tight time frames. And I must confess, I was woefully unprepared.

And in the constant struggle to generate fresh material under such enormous stress, I began to notice something curious; namely, that the best material seemed to occur to me when *I was either driving or taking a shower.* Sound familiar?

Though grateful for these helpful revelations, I initially dismissed this as a coincidental quirk. Neither of these activities *seemed* like they should be particularly productive. After all, they appeared to be the *opposite of actual work,* which I—and my employers—presumed was supposed to happen at my office computer, or someplace equally suited to the task.

And yet, the material continued to 'bubble up' while in the driver's seat of my VW hatchback, or my morning hosing, with consistency. I soon found myself making excuses at the office to jump into my car, or rush to the nearest gym for a quick shower whenever I was under a deadline, and was hard pressed to crack another story or untangle a seemingly impossible plot twist.

Of course, constantly disappearing from work on vague "errands" and often returning freshly showered inevitably led to rampant speculation around the office: I was either on drugs, or a shameless lothario. Both possibilities were tolerated in Hollywood as long as I continued to deliver usable scripts on schedule. Showbiz; you gotta love it.

A more meaningful result was the inescapable question that occurred to me: *What was really happening when I drove around or showered* that not only allowed, but also seemed to encourage such useful solutions to come to mind? And might there not be a way I could get the same thing to happen without hopping in the car? Or running around Hollywood looking for a shower?

At that point, I didn't really understand what was at the heart of my curious creative process. Such a subtle but complex question might've seemed daunting, if not beyond my reach entirely. But as I was regularly weaving intricate stories of plot and character — and doing this very successfully — I figured maybe I could make some sense of this peculiar pattern. And, I was determined to get to the bottom of it, one way or the other.

The MacGyver Secret

So over the next few years — and through subsequent staff and other writing jobs—I began to explore this creative process, and experiment with various techniques to see if there was some way to better understand what was really happening in me, and how I could capture the effect of driving or taking a shower in a more convenient and efficient way (gas and water being somewhat precious resources).

And, as fate would have it, it was when I was writing the pilot script for MacGyver — that master of ingenuity — that the pieces of this puzzle finally came together in a complete picture, and revealed the essence of this mysterious process. Along with it came a remarkably simple and workable technique for producing my driving/shower epiphanies. I soon learned to produce good ideas virtually on *demand* in nearly any setting I chose. More about the details of that soon.

But the bottom line was, I had uncovered the secret! I had found my *inner* MacGyver.

Now, to be honest, I didn't really think of it like this back then. I mean, who knew if the MacGyver pilot would even be made, much less turn into a hit series, a global phenomenon, or such a popular verb that it was ultimately even recognized by the Oxford English Dictionary!? ("To MacGyver"– Go ahead, look it up.)

Nor did I imagine it had any use or applications beyond my own creative process, or that it might work for anyone else. I was, after all, just another Hollywood TV writer. This is not a group exactly renowned for its intellectual breakthroughs.

But, I knew I had just cracked the secret: I found the most amazing creative process that never *failed* to produce great results, wherever and whenever I needed them.

And, as I refined this secret over the ensuing three plus decades of my career, the liberating impact of it both personally and professionally proved nothing short of *transformational*.

The stress of creation and problem solving was all but *eliminated*. And, knowing I had access to the extraordinary resources of my *inner* MacGyver rendered me fearless in the face of any project or deadline.

Over time, I would casually share this secret with friends and acquaintances, most of whom found it too incredible to take seriously. After all, it's nothing like the way *we're taught to solve problems*. Fair enough.

That is, until I offered it to a young friend and colleague, Jared (who will tell you his own story on page 36). Jared had landed a job to launch an internet company, and he decided to try using the secret to solve the constant crush of problems that comes with such an undertaking.

Well, having experienced the same transformational results, Jared came back to me raving about how amazing and effective it was. He argued that this secret could be successfully applied to any number of endeavors, well beyond the creative writing I'd been doing with it.

Was it possible I had somehow stumbled upon the secret to unlocking the "Swiss Army Knife" of the mind?

Much discussion with other friends and colleagues followed, along with a bunch of research into the current state of the science behind this secret. I was eventually persuaded that this is indeed a secret to share with anyone who wants to create great ideas and solve tough problems.

I mean, WWMD? (What would MacGyver do?)

Chapter 3

How I Put My Inner MacGyver to Work

So, having reached the obvious conclusion that I could neither drive around nor shower in my office, here's what I did. I installed a white board and a workbench in there instead. The white board was for my questions– and answers. The workbench was to *build models*.

I mean, my bosses are right outside my door, having just told me, "The network rejected the story we were planning for the next episode. We need a new one fast, Lee. What have you got?" No pressure, right?

So I'd write on the empty whiteboard "What's a great idea for an episode?" And then rather than wrack my brain, I would say to my inner MacGyver—sometimes silently, sometimes out loud—"You're the one with all the good ideas. You come up with a new episode for me." And that, in an hour or so, *I'd be back for the answer.*

Then I'd go to my workbench and set to work on a model. And do my best **not to think about the question... at all.** I'd just focus my thoughts and hands on building the model. I started with paper ones of famous buildings. Like The Empire State Building or The Vatican or The Taj Mahal.

In time, I built every kit I could get my hands on. Cutting, folding, gluing, sticking Tab A into Slot B fifty times, and so on. A seemingly 'mindless' activity. But it had the exact same effect as driving or taking a shower. Namely, keeping that endless hamster wheel of spinning thoughts in my head **occupied** so *my inner Mac could work on the question* and, frankly, leave 'me' out of it.

After about an hour, I'd then set the model aside and return to the white board, pick up a marker and say "Okay, Mac, what have you got for me?" And then I'd just start writing. Anything. It could be what I wanted for lunch or the lyrics to a song. It didn't matter *what* I started writing. Because, in about thirty seconds of that writing, ideas for new episodes would just start *popping into my head!*

For example, "What if an old friend comes to our hero because he's in a jam? But it turns out it's really a set up to frame our hero for some bigger crime? Wait—what if the old friend is instead an old flame? And maybe they're in on the set up? But maybe they're not?"

The ideas often coming in faster than I could get them down. And after no more than five or ten minutes of furious transcribing...*there they were.* Exactly what I'd asked for: At least one or more killer notions for a new episode or story. Just like that. My inner MacGyver had done it again.

I can hardly describe the thrilling rush of watching those ideas spring to life, *effortlessly.* Not to mention the utter relief of knowing I now had something really solid to work on in the face of another looming deadline.

So I'd bask in that giddy feeling for a few minutes, letting the ideas dance around in my head as I decided which way I wanted the ideas to play out. Maybe one would be the main story—what us TV hacks call 'the A story'– and one would be the B story. Like, in the example above, re-playing the original romance or whatever. And once I'd settled on those details then...I'd ask the next question. **And do it all again!**

In this case the next question might be "What's the best structure for this story?" Or, more specifically, "What are my act breaks?" Because, in the network shows I wrote for, each episode was divided into four or five 'acts', separated by the commercials. And you generally needed some form of story turn or 'cliffhanger' at the end of each act to keep the viewer hooked between the commercial breaks.

So I'd set my inner Mac to puzzle out the act breaks, with another time limit, and return to whatever paper monument I was working on.

And I'd do that repeatedly. All day. By the end of the day I might've spent six or more hours working on the model and maybe an *hour or two at most* at the white board. Sounds crazy, I know. Except for the fact that it would usually take a writer at least a week if not more to lay out a whole story. And, using my inner Mac, I could crank out an entire story—scene by scene—*in a single day.* Without ever having to 'wrack my brain' once. Or break a sweat.

What's more, the questions could be incredibly detailed and specific. Like 'I need a scene that's a little bit funny, a little bit scary, reveals something about the characters, and advances the story with an unexpected twist that still makes perfect sense.' And so on. Nothing phased my inner Mac. Because, after another cycle at the whiteboard and time working on a model, Mac would come back with a whole set of options—all of which met *every request on my list.* Scout's honor.

That meant I could use my inner Mac clear through the writing of the script. Tapping into it whenever the words stopped flowing or I thought the piece needed another injection of something cool or more original.

Now you don't need a whiteboard. Or to build models, for that matter. Any writing implement and piece of paper will suffice. And there are literally hundreds of other activities you can do to divert the hamster wheel in your head.

But now that you have some small sense of how the secret works—at least for me—let's talk about *how the MacGyver Secret can work for you.* Because, believe me, your inner MacGyver is just waiting for you to turn it loose on a problem!

Chapter 4

How The MacGyver Secret Can Work for You

Now that you're at least curious—and adventurous—enough to really look for your inner MacGyver, a bit of a roadmap might come in handy; because, as easy as it is, this little journey is probably unlike any other you've embarked upon. And may well take you to places that the GPS on your phone can *never* find for you. Imagine that.

For instance, you might've already noticed the little Swiss Army knife and lightbulb icons atop the last few chapters? Good catch. Because those are some of the signposts to guide you through the book, so you'll always know where you are and help take you right where you want to go.

 When you see this little Swiss Army Knife, it means you're on one of the **instruction steps** of the MacGyver Secret. These explain all you need to simply and easily connect with your inner MacGyver. Along with some Practice Problems if you'd like to take your inner Mac out for a test-drive before turning it loose on a question or problem of your own. Not to mention every trick and tip I've learned from using the secret for decades and teaching it to hundreds of others, just like you. There'll also be a short list of bullet points at the end of those chapters, summarizing all the key points. Plus, a Cheat Sheet at the end that has them all together in one place. So, if you're eager to roll up your sleeves, and really want to dive into the heart of things as quickly as possible, then just look for that little knife and you'll be good to go.

 As you've most likely figured out already, the lightbulb icon is for the ***stories*** in the book. Some are mine, some are by entrepreneurs, designers, writers and others who I've taught the secret to, and who have been using it in their work and lives for years. Some are by the people who've encouraged and helped me to share the secret with you here. And some are about famous inventors, artists, scientists and others whose stories of their breakthroughs echo and validate the key elements of the MacGyver Secret. So, when you see the lightbulb, you'll know it's story time.

 This icon, of the brain and paper clip, (one of my favorites, obviously) is your key to the ***science*** pieces in the book. No doubt you also noticed Dr. Colleen Seifert's name as my co-author on the book. Colleen is a respected scientist and academic with an impressive resume, including a Ph.D. in Psychology from Yale and a full professorship in the Psychology department at the University of Michigan. Colleen now teaches and conducts research there on cognitive science, memory and a host of other related subjects– when she isn't running studies to confirm the efficacy of the MacGyver Secret. Not too shabby, eh? Her full bio is at the back of the book and you can also find her own story of how she became part of the secret on page 46. Suffice it to say I am humbled and grateful to have such a brilliant and acknowledged authority writing all the science sections here that explain exactly how and why the MacGyver Secret works. So, if science is your thing– or you're just curious about the research behind all this– then look for the little brain so you can plug all that info right into yours.

You guessed it. The little hand and pen icon means it time for YOU to jump in and **write** something to get that crucial chat started with your inner MacGyver. Or make a list of some things to customize the secret to fit exactly the way you live and work. We've even left you enough space so you can do that right here in the book if you want. And some of that precious space is, of course, to write your answers. To either the Practice problems or YOUR OWN questions— which is the real pot of gold at the end of this particular rainbow. Because the answers are there in you! And once you see how easy it is to connect with your inner MacGyver, they'll be all yours for the asking. So when you see that little hand icon, don't be shy. Grab that pen and go for it. Because that's your key to unlock any problem you're facing.

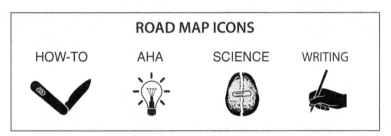

To go straight to the science or the 'aha' stories, there's an index on page 145.

Still not convinced? Just sounds too good to be true? I hear you. That's what everyone says. *Until they try it.* And that's when the story takes an amazing turn and suddenly...is a story no more. But instead becomes a whole new reality! And I don't have to say another word.

Because that's when people just like you start telling me *their stories* about all the incredible breakthroughs they're having. Again and again.

And now it's your turn. So hop on board and see just how far this journey can take you. You can read everything, or use the icons to guide you through the book in whatever way or at whatever pace works best for you.

If you have any additional questions or experiences you think are worth sharing, feel free to make a pit stop at our MacGyverSecret.com website. We'd love to hear from you. We have a YOUR VOICE page all set up just for that. You may discover others have already been at that crossroads and can point you in the right direction. And if they can't then rest assured, we will.

That's all there is to it. *The secret is now yours.*

My only hope is that, should you decide to try it, it will change your life as much as it has changed mine.

THE SCIENCE: WHAT LEADS TO 'AHA'?

Almost everyone has experienced *insight;* you give up on a difficult problem, and then later, the solution suddenly comes to you: 'Aha!' Just like Lee, you may have wondered why these breakthroughs come in the shower, and not when you need them at work! The answer may lie in relaxing within the stream of our daily experiences.

Studies of brain activity during problem solving have identified a resting-state brain activity marked by diffuse (rather than focused) attention.[1] In this relaxed state, we may be better able to allow unusual associations to emerge into consciousness.[2] This "default" mode of thought appears to support our ability to daydream so effortlessly (and often: We spend almost 50% of our days thinking about something other than what we're doing[3]).

Having an insight may depend upon disengaging from the external environment. Turning your attention inward -- such as closing your eyes or staring blankly without "seeing" -- may allow more focus on an internal train of thought.[4] In the lab, eye-tracking measures show that people are more likely to blink and look away just before an insight occurs.[5]

Interestingly, the mind is only sometimes aware of its wandering. Brain imaging studies show that the default network and executive control systems are even more active during 'zoning out,' when we're not even aware that our minds are wandering.[6]

Getting "lost" in driving or showering, as Lee describes, may help to prepare our minds for new ideas to arise.

The MacGyver Secret

PART 2

The Three Steps of The MacGyver Secret

The MacGyver Secret

STEP ONE

Chapter 5
Getting Started

As you could see from how I did it, there are really only a few simple steps to accessing your inner MacGyver because, rest assured, it's already there in you and ready to help with whatever you need.

All you need to do is ask. That's the first step. Boom. *Just ask.*

But to really wake up your MacGyver, you can't merely think about your question or problem, or even say it out loud. Just as I did with my whiteboard:

You need to write it down.

That's all it takes to dial in the "super solver" inside you. And, while it's okay to type your question or problem into a computer or tablet or phone, since Mac is very much a *hands-on* character, it works even better and faster if–

You write the question or problem down in longhand.

Any pen or pencil will do, as will any piece of paper or whatever– or the blank pages we've set aside for you at the end of the chapter. Though, you might want to consider finding a notebook sometime soon to keep all your questions—and answers—in the same place. But for now, just grab any writing implement that's handy and you're all set.

You're wondering: "That's it? Just write my question or problem down? In longhand? *Why?*"

Short answer: Writing it down makes your question or problem *clear and concrete*. It makes you process the question deeply, and makes it your own personal questions. And it signals your inner MacGyver that this is *really what you need some help with,* as opposed to all the other thoughts, fears, questions and noise that generally run around in your mind all day. Even if all it appears to be is a bit of graphite or ink on a page, you've now given your question or problem *actual substance.* And made it *real.* And that matters, a lot.

As for why it's better to do it longhand, I'll let Colleen walk you through the science on that if you're interested later in this chapter. But the bottom line is, *typing is not the same as writing longhand.* Seems like it shouldn't really make any difference. But it does. And, to really tap into your inner Mac, old school works best.

In the next chapter, we'll give you some tricks and tips on how to really hone the asking step to a fine edge. But Mac has always been more about doing than talking, so let's just *get you started with opening the door to your inner MacGyver.*

The MacGyver Secret | Step One

So, **right now** write down a question or problem that you'd like some help with. It could be a creative problem, or technical, or personal, or *anything*. Just write it down **NOW**. It could be one line, or a whole page in great detail. Doesn't matter. And don't worry about the answer. That's for your inner MacGyver to deal with, remember? **JUST DO IT.** Take your time. We're not going anywhere.

Still not sure what to ask? No worries. We have some practice problems for you right after this if you'd like to start with those. Sort of like warm-up exercises to get your inner Mac off the couch and jogging down the block.

The MacGyver Secret | *Step One*

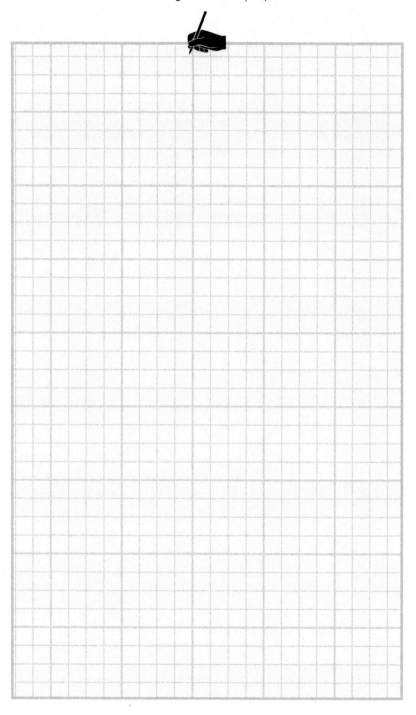

PRACTICE PROBLEMS

Here are a number of different types of problems you can use to practice the MacGyver Secret until you're ready to try it with questions or problems of your own. You can try them one at a time, or several together when you have the hang of it. All of these problems are simple and open-ended, meaning they have a lot of possible answers. They're also non-obvious types of problems which means they require some thought.

First, read through a problem. If a great answer comes immediately to mind, skip it and try the next one.

Then, **write the problem down either here in the book or on a separate piece of paper.** Once you've done that, then just go on to the next page and we'll tell you what to do next. Remember, you're just writing the question *so don't give another thought to the answer.* Because that's for your inner MacGyver to solve.

1. Choose a close friend or relative, and come up with an incredible gift to get them for their next birthday.

2. Write a short story in the form of a voice mail message.

3. Come up with a story about a person who finds something he or she has no intention of returning.

4. Think of as many different uses for a ping-pong ball as you can. Be creative and practical.

5. Think of a great way to spend a vacation week next summer, including "who, what, when, and where."

6. Come up with a new activity you can do to spend an afternoon with your child, parent, or close friend.

7. Come up with ideas for the "Best New Unappetizing Theme Restaurant."

8. Pick a specific experience you had where you failed, and come up with ways to turn it into a celebration.

9. Make a wish for a creative goal you want to achieve this year, including steps to make it come true.

10. Come up with ways to get people to stop using disposable straws.

11. Think of a great outing you can do with your friend who is single to help them meet other single people.

12. Identify your personal brand. If you were to place those attributes and characteristics into a wine bottle, what would you name it and what would your label design include?

13. Think of 3 random acts of kindness that you can do later today.

The MacGyver Secret | Step One

Good! Now that you've either written down your own question, or written down one of the practice problems, here's what you do:

Tell your inner MacGyver to work on it for you!

You can do this silently or out loud (depending on how you feel about talking to yourself.) But either way, *it's important to do this.*

If you had a chance to see the original TV series, you know that MacGyver often spoke to himself– so that you could hear his thoughts even though he wasn't speaking out loud– with what we in the entertainment biz call *Voice Over;* hearing his inner thoughts when he was trying to work out a problem and gathering random things to construct a solution. Well, you are *now trying to build a relationship with that same awesome problem-solver deep inside you.* And saying 'hello' to it– and **telling it** that you need some help– is the best way to set it in motion even if it feels a little weird. This is MacGyver we're talking about after all. So a little weirdness can't hurt, right?

Okay, so you've written down your question— (or one of the practice problems). And you've told your inner Mac to have a whack at it. Now what?

Just put it aside **and forget it for now.** And move on to the next chapter. Because, even though it may not seem like it, you've just thrown the switch to turn on your inner MacGyver. And help is already on the way.

THE SCIENCE: WHY DOES WRITING IN LONGHAND HELP?

Writing out your thoughts likely serves several purposes. Describing your problem in writing forces you to understand it more deeply, clarifying a rambling stream of thoughts into an economical form.[1] In writing it down, you edit your thoughts to come up with a more careful expression of what you really want.

Writing your problem also allows you to return to see it exactly as you wrote it when you're ready to call for an answer. You're more likely to retrieve information from memory when the circumstances at encoding (when the information goes into memory) match those during retrieval.[2] So when it's there on paper when you "call for the answer," you can see again exactly what you asked for. Try to keep the setting where you ask your inner MacGyver for help the same one you use when you call for the answers.

There may also be something special about the act of handwriting (or printing): The physical act of writing is a more distinctive, embodied process than typing.[3] The messiness of free-form handwriting requires that we plan and execute the action (and deal with a result) that is highly variable. Producing a messy letter may actually help in remembering it. Imagine writing your problem on a whiteboard (as Lee does); in that case, your whole body becomes engaged in the act of writing, which may help your mental processing.

Finally, writing may help to break the conversational ice with your inner Mac. Talking to yourself can be an awkward conversation at first! Writing to yourself (your inner Mac) is a form of *self-distancing*. By referring to yourself just as you would to another person, you create a psychological distance,[4] especially helpful when under stress or "on the spot" (such as facing a deadline for a script to be done). This small shift in the use of your language has been shown to help people reflect on their thoughts more objectively.[4]

Shifting from 'I' to "you" can also help you to accept your thoughts and feelings.[5] So instead of saying, "What ideas have *I come up with?*" you can say, "What answers do you have for *me*, Mac?" and be more open to the results. Lee uses this language extensively when drawing out his ideas, and doing so may play an important role in his success.

JARED'S STORY

You might remember I told you that I was encouraged to share the MacGyver Secret because it helped my friend Jared transform the way he innovates as a Tech Entrepreneur. Well, this is his story, in his own words.

As a kid, I was a MacGyver devotee. I mean, I had the alarm on my 1980's Casio wrist watch set to go off five minutes before MacGyver came on every afternoon in re-runs for years so I wouldn't miss a single episode.

After graduating from film school, my Dad casually mentioned that he had met the creator of MacGyver socially and could probably arrange a meeting if I was interested. Interested?! Honestly, I was ecstatic at the opportunity to meet Lee.

That meeting turned out to be the first of many, and the beginning of a lifelong friendship. But the most unexpected part was learning how Lee had come up with the MacGyver Secret. As an aspiring young filmmaker, I was always on the lookout for secrets and tips to help me "break in" to the film and TV industry. This looked like my big chance.

Over the course of that meeting, and others, Lee outlined the secret he used to generate new story material. It sounded awesome and simple enough. And it was coming from the creator of MacGyver! But maybe too good to be true. So, of course, I promptly forgot about it for the next five years.

Cut to five years later. I'm now running my first tech company, and really, really stuck trying to figure out the business model. I met Lee for dinner and he once again suggested that I try his secret. But this time, knowing how confused and desperate I was, I decided to give it a shot.

The MacGyver Secret | Step One

So the next morning, I wrote the problem down, went for a bike ride, came back to the paper and...nothing. Useless, I thought. But then, after a few seconds, this one idea suddenly came to mind. And then another. And then I saw how those two ideas could come together.

I was a little incredulous that the secret had somehow created those ideas, but I was intrigued. Next time I was stuck, I tried again. And it worked! Again. And again, and again, and again. Some of the best ideas I've ever had. Ideas that saved my company many times. Ideas that shaped my career. Whenever I asked. On demand.

My office at the time was near Venice beach. And nearly every morning, I'd come in to work, write my questions on a whiteboard or a yellow pad, and then go for a bike ride along the Venice boardwalk. Or, if the waves were good, I'd go body surfing—for work! Then I'd come back, return to my questions and the answers would just come— every time!

After a few years like that, I realized that it was crazy for Lee and I to be the only two people in the world using this secret. Everyone gets stuck, needs to solve problems and searches for insight into their situation. Why should Lee and I be the only two with this amazing, built in, on demand system? So I went to Lee with a simple message: you need to share this secret with the world.

It took some convincing, but since I was using the MacGyver Secret to solve problems that had nothing to do with stories or scripts, I finally got Lee to see its potential. For entrepreneurs, engineers, people looking for personal growth, for anyone.

And I still use the secret, all the time. Mainly, when the path forward isn't clear. I'm frequently confronting decisions and situations with very little info about outcomes, but a requirement to make a call and move my team forward. The secret helps me get clear and produce insights into the decisions that are otherwise hard to come by.

I also use it for solving product problems. A big part of my job is developing solutions to specific user or business problems at the product level. Very often the solutions are non-obvious and I'm under a deadline. But, with the secret, I create solutions more quickly, and with less stress than any other system or process I've come across.

Honestly, the biggest challenge I face with the secret is remembering to use it! It's so easy to get caught up in the day to day crunch of stuff that I forget it's right there in my toolbox. But, I will tell you straight, every time I reach for it, it works!

Jared Krause
Technology Entrepreneur

Chapter 6

How to Ask the Right Questions - Or Asking Questions Rightly

Asking questions is usually no big thing, right? Like, 'What time is it?' or 'How much does that cost?' But most of us generally don't ask **ourselves** a lot of questions beyond things like, 'Do I want a medium coffee or a large?' or 'Should I wear the blue dress or the green slacks?' And even fewer of us have any experience asking questions or posing problems to *our inner MacGyver.*

So, it might be worth looking a bit more closely at that part of the process so you can really get the most out of your inner Mac. I mean, if MacGyver is standing in your kitchen, you'd really want to take advantage of all his talents, wouldn't you? Rather than, say, asking him to just change a lightbulb.

Not that asking questions of your inner Mac is in any way hard, because it isn't. But since you've just met, some pointers may come in handy. And we've got some sample questions here for you too to make it even easier.

Now, anyone adept at solving problems will surely tell you that the first step is to do your best to *try and define the problem.* As clearly, and in as much detail as possible. And it's no different with your inner Mac.

So, when writing down your question or problem, **take your time.** And try and be as **specific as you can.** Even if that means you end up writing several paragraphs or a *whole page*. Or pose the same question or problem in different ways. Let's face it, the more details MacGyver—or your inner Mac—has about a situation, the more likely it'll be able to come up with an awesome solution, no?

Take it from me, you simply can't give your inner Mac *too much information*. Like, for example, when I'm looking for just the right scene in a script, I write down a whole *laundry list* of specifics that I want. Or I might write out a long list of 10 or more different questions, like 'How many of my characters do I want in this scene? Who begins the scene? Is there a shift in the scene between characters? What do each of the characters want from one another in the scene? How does it build? How does it end?' And so on.

Mac can handle it all. Promise.

That doesn't mean you can't ask broad or general questions, like 'Why can't I find a job I really like?', or 'What's it really gonna take for me to lose 20 pounds?' But, as you might expect, the more general or vague the question, the more likely your inner MacGyver is to respond with a broad or general answer.

And don't be surprised if your inner Mac answers you *with more questions*. It happens all the time. So if you ask 'Why can't I find a job I really like?' your inner Mac may come back with, 'What do you think you're good at? What do you really like to do? Who can you think of who gets paid to do *that?*'

In which case, it's totally cool to then *re-write those questions* and throw them right back at Mac for some specific answers.

Because the thing here is you're **really starting a conversation or DIALOG with your inner Mac.** Bouncing questions or problems back and forth—however frequently you choose to do it– once a day, once every other day or whatever– until you get the answers you can really use. Like when I'm trying to create a new story, I might go back and forth between asking questions and working on a model— or a fancy meal– five or six times a day!

Your inner Mac *will never get tired or fed up with your asking.* It's **YOUR** inner MacGyver after all, and is there to pitch in and help however often or badly you might need or want it.

One more tip. If, when you're writing down your questions or problem, some quick answers pop into your head, **then write those down too!**

Chances are your inner Mac has even better and more useful answers for you. But you don't want those first, easy answers rattling around in your head while you're doing your activity. So write them down and *just put them aside with your questions.* Your inner Mac will let you know if they're really the solution you're looking for.

REFINING YOUR QUESTIONS

So, before we send you back to look at the question or practice problem you chose earlier, take a moment to look at some of these examples of questions. You might find it helpful to re-write them, filling in the blanks with whatever you'd like your inner MacGyver to work on.

1. What is it I really need to answer about _____?

2. What answers have I been considering about _____?

3. What are the key questions I should be asking about _____?

4. What are some possible answers to _____ that I haven't considered?

5. What is at the core of the problem with _____?

6. Is _____ really the problem or could it be something else?

7. Have I really considered all the aspects of_____ yet, or am I missing something?

8. If deciding about _____ is just an either/or question, what are the real key factors behind that decision? And are there some I haven't yet consiered?

9. Now that I listed everything about _____ that I can think of, is there one perfect answer? Or might there be a number of possible solutions to it?

10. Now that I have a number of possible answers to _____, what do I need to really choose the best one?

The MacGyver Secret | *Step One*

Okay, so if you haven't already gone back and taken a look at the question or problem you first wrote down, **do it now.** And see if you can't expand or revise it either there or here to really give your inner Mac something to sink its teeth into. And don't worry, your inner MacGyver will pick up any changes you made and keep right on humming along.

The MacGyver Secret | *Step One*

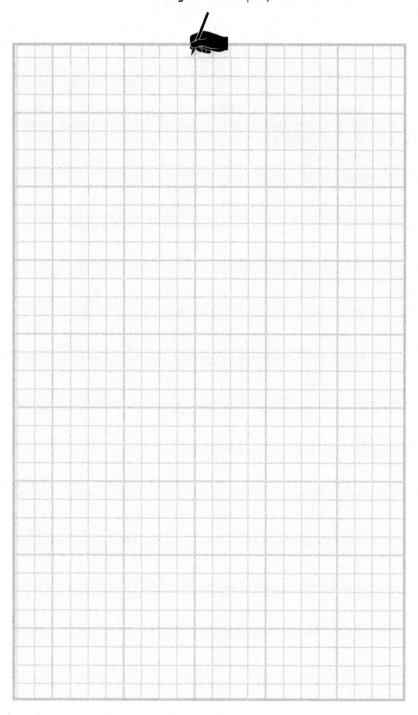

THE SCIENCE: DEFINING THE PROBLEM

In our school experiences, our thinking often starts with a presented question. We learn to solve problems, analyze causes, and derive equations, but we are never trained in the art of posing the problem! In life, we frequently experience the frustration of knowing a problem exists, but have trouble identifying exactly what the problem is.

For example, suppose your friend Zeke is always late. Is the problem that he does not have a watch, or that he can't remember when he was supposed to arrive? Or is he expressing his dominance over you by making you wait for him? Each of these views suggests different possible solutions! So in order to pose it clearly, you need to 'discover' your problem.

Problem finding refers to the ability to construct a problem, identifying its key elements and formulating an apt description.[1] Problems must be deeply analyzed in a process called *immersion* before a solution can be achieved.[2] A study of artist students found that those who spent time discovering and refining the problem initially before actually drawing the image (e.g., a 'still life' scene) produced works rated as more creative by experts, and even enjoyed greater success in the field fifteen years later.[3]

Studies have also uncovered evidence that problems and solutions co-evolve, where discoveries made in trying to solve a problem are used to refine the problem itself.[4] This cyclical process of problem and solution discovery is suggested by Lee's process as well: If you're not sure what the problem is, then that is yet another problem to consider!

COLLEEN'S STORY

In case you were wondering why someone with Colleen's sterling credentials would decide to cast in with likes of a Hollywood hack like me, here's that story in her own words. I always figured it was just because I was a good talker. Guess that wasn't it.

In June of 2012 I received an email out of the blue. It was from Lee, who had been referred to me by a mutual friend and cognitive scientist. Though a Hollywood writer, Lee explained that he was working on a project about creativity and the mind/brain processes involved. He was interested in learning more about the current state of scientific research on the subject, and how his creativity project might fit with all that. Would I perhaps be open to a phone call to discuss this with him?

As a professor of psychology at the University of Michigan in Ann Arbor, I've had my share of odd encounters with people who wanted their "special questions" answered. One involved a gift of a green Jello mold left at the office door. But when this email came in, I shot up in my chair. As the creator of MacGyver, among dozens of other credits, this Zlotoff fellow looked like the real deal! I wrote back the same day.

In the ensuing phone calls and emails, we began our conversations to build a bridge connecting what we know from the world of science to the MacGyver Secret. As Lee laid out his experiences with his creative process, several things were immediately apparent. First, he is highly intelligent, and had a well-educated intellect. He clearly described his solution to the problem of constant demands for creativity. He was also intuitive and insightful in observing his own thoughts and behaviors, and in identifying the important elements of the process he had discovered.

Immediately, it struck me that Lee was on to something... As a researcher working on problem solving and creativity, this was right up my alley. I could see where the science was consistent with his 'secret'; and, that his thinking could advance the science by pushing on questions that are too "hairy" to address in the lab. First, Lee describes a complete creative process, while scientific studies address only a small portion of the process using artificial laboratory tasks. In addition, Lee's experiences involve much more complex problems, unique creative solutions, a time scale of mastery over years, and set in real-world, high pressure creative environments. He offered me a quite a challenge!

I jumped on board. That March, Lee came to Michigan, and we held a workshop with a group of my colleagues, students, and friends. It included a choreographer, a product designer, and one of the world's experts on memory and unconscious processes. While they had helpful critiques, Lee's presentation of his 'secret' passed their test as a novel and interesting take on the creative process.

From there, Lee and I interacted several times each month by email, as incredibly exciting people and events flowed through our lives (well, Lee's). Lee gave some talks at businesses and universities, along with more workshops. We refined the message over time, adding in the language of cognitive psychology to clarify the points Lee wanted to make. I've encountered others who talked about similar experiences while creating, such as relying on incubation. However, Lee's MacGyver Secret is unique because it provides explicit instruction on *how* to engage in this creative process.

A highlight for me was a week-long virtual workshop with professional writers. I was able to listen in as they talked about their experiences, their ease in using Lee's process, and their genuine awe at finding that this secret worked for them, too! I began testing parts of Lee's process with laboratory experiments and classroom exercises. My students greeted the MacGyver Secret with real enthusiasm, especially when they were asked to incubate during class time. And the results were very promising. Many talked about continuing to use Lee's process in their own creative work even after the course ended.

Four years after that first email, Lee finally found the time from his own creative work to write this book. Adding in the science supporting his claims was exciting for me because it ties together many separate ideas within the field. It's astonishing to see the connections between Lee's descriptions of his secret, and the latest findings on insight, unconscious processes, and creativity. Some of this research did not yet exist when Lee and I began our talks. Many streams of research support it, and I'm not aware of any existing findings that are inconsistent with the MacGyver Secret.

A final challenge is that Lee's creative process was devised, practiced, and mastered over a period of years. This scale of effort falls far outside the current scientific protocols for experiments on creativity. Though I have dreamed of it (literally), it has not yet been possible to do a full, randomized controlled trial of the MacGyver Secret's effectiveness. Perhaps, as a result of this book, such a study will be possible. Until then, however, we are open and eager for feedback from you to determine whether this process can be mastered by more creators and problem solvers. So, for the sake of science, we invite you to share your experiences at MacGyverSecret.com.

Colleen Seifert, Ph.D.

STEP ONE: SUMMARY

Okay, to give you a quick review then, here are the bullet points for Step One:

- Write down your question or problem *(longhand is best)*

- Take your time.

- Be as detailed and specific as you can.

- Don't sweat about overloading your inner Mac.

- Write down any quick answers that occur to you.

- Tell your inner MacGyver to work on it for you.

- Then *let it go!*

Piece of cake, right? And, now it's time to move on to STEP TWO. Which, believe it or not, is *even more fun and easier than the first step!* Seriously? Read on.

The MacGyver Secret

STEP TWO

Chapter 7
Letting Your Inner MacGyver Get to Work

So now that you've written down your question or problem, and told your inner Mac to work on it, what should you do next? Simply put, *you should just stay out of the way and **let your inner MacGyver do its thing.***

'And how exactly do I do that?' you're wondering. Good question.

Imagine for a moment you were caught in a tough situation with MacGyver. What do you think he'd say if you kept pestering him with an endless stream of questions: 'What are we gonna do now, Mac? Do you have a plan? Any ideas, Mac? What do you think we can use to get us out of here?' And so on. Most likely, he'd politely suggest you put a cork in it so he can, you know, think clearly and find a good way to deal with the problem.

Well, it's no different with your inner MacGyver. You need to find some *activity*—preferably an enjoyable one—to keep all that noise in your head preoccupied with *something* so that your inner Mac can have the time and space it needs to solve your problem. (Colleen's term for these are 'incubation activities' which she'll explain in the next science section).

In my case, it was building those paper scale models. Not because I really needed a desktop rendition of the Brooklyn Bridge. Who does? But because it was the perfect way to keep my *mind* occupied so I could really put the problem aside—or forget it entirely– to let my inner Mac do its thing.

And building models required just enough of my attention that it kept me *focused* on something other than the problem I'd asked Mac to crack for me. It kept me preoccupied so I could stay out of the way of my inner MacGyver. Just like driving and showering.

You have to pay attention when you're driving, no? Even though it may be kind of second nature, you still have to be aware of the cars in front, back and to the sides of you. And the traffic lights, and the route you're taking. Second by second.

The same is true in the shower. It's kind of automatic but you don't want to get soap in your eyes, or water up your nose, or slip and fall. Both tasks require *constant monitoring* to adjust your actions based on what's happening. That helps to keep your hamster wheel—or conscious—mind off the problem you gave to Mac. But neither task is very challenging, and so they leave plenty of mental energy for your inner Mac to work on the *real* problem.

But if building models isn't your thing, not to worry. There is no shortage of things you can do that will work just as well.

Like doing a jigsaw puzzle, or cooking, or gardening, or knitting, or practicing a musical instrument, just to name a few. The important thing here is that, in addition to demanding your focus, it should have some **physical** aspect to it. Even if that's only moving your fingers to fill in the answers in a crossword, Sudoku or word search puzzle.

And if that's not physical enough for you, then you can try going for a walk, or a jog, or a bike ride, or a swim, or shooting some baskets, or working out at the gym or on that stationary bike that's been collecting dust in your den. Or nearly any sports activity that isn't so *demanding* that even your inner Mac had better be paying attention, like, say, mountain climbing or hang gliding.

Not to mention the classic standbys of going for a drive or taking a shower. Both of which you might notice also have a *physical* aspect to them, be it steering and stepping on the gas or shampooing your hair.

So even though your inner MacGyver is doing the heavy mental lifting, your body needs to be involved—if only in subtle or minor ways—to keep the hamster wheel in your mind focused elsewhere.

But what if I don't have the time to do some enjoyable activity?' you might be thinking. Again, no problem. Just do some *routine activity you have to do anyway*. Like cleaning the house, washing the dishes, walking the dog, organizing your clothes closet or sock drawer, or any semi-mindless activity, which requires that you will need to get it done sooner or later anyway.

Though I started with building models, these days I find cooking and doing small remodeling projects around the house work best for me. And from the research and workshops we've done, it's clear that the effectiveness of what Colleen calls these 'low-demand incubation activities' varies from person to person.

Learning to draw dinosaurs, or filling in adult coloring books say, was ideal for some people. But doing a crossword or a jigsaw puzzle was just frustrating. While others found doing Origami, or building with Legos worked best for them.

So don't be afraid to experiment with a whole range of activities until you find one or two that you know will keep your mind focused enough that it simply *can't or won't* think about the question you've posed for your inner MacGyver.

And, while it may *feel like* none of these activities is productive, the fact is that your inner Mac is quietly *working the whole time* to solve your problem.

'Hang on—', I can hear some of you saying, 'I work in an office all day, 9 to 5 if not longer. How am I supposed to do *any* of these activities at work?' Mmmm. You must've missed the word *MacGyver* in the title, huh? No matter because, yes, we've even MacGyvered a way to make the secret work in an office or other severely restricted setting. And we have a whole chapter just for you: *The MacGyver Secret In The Workplace*. So relax, we've got you covered.

Next question: Are there any activities that **don't work** for my inner MacGyver to do its thing? Yes, there are. Just a few. But they're biggies so brace yourself.

Watching TV won't work. So, no YouTube or Facebook videos. Neither will **reading**, a book, a magazine or anything else. Nor will **conversation**, whether talking, listening, extended texting or chat, or conversing over email. And last, but not least, **playing intensely interactive video or computer games won't work**. Simple digital games are okay, Candy Crush, Tetris, Angry Birds, and the like. But I'm afraid **first-person shooter or role playing games are out**: World of Warcraft, Call Of Duty or any game that involves you going into and interacting with another world.

But why?! Colleen will give you the scientific reasons why those 'high-demand' activities prevent your inner Mac from doing its thing in her next piece. But the short answer is this: *All* of those activities—even the ones that seem *totally passive,* like watching TV or reading— actually need a lot of your inner Mac's bandwidth to make them possible. Figuring out what something *means* takes all of your inner Mac's attention. And if it's busy doing that for you, *then it can't be solving your problem!* Granted, MacGyver is awesome. But not Superman.

So if you really want its help, then lay off the TV, the Playstation, the books and yakking on your phone. It's fine to take a quick call or answer a co-worker's question. But try to refrain from extended conversations if you can. Just saying.

'What about taking a nap or sleeping? Can my inner Mac do its thing if I'm asleep? As a matter of fact, *it can!* Because, even though your hamster wheel mind needs to hit the pillow every night to stay functional, believe it or not, **your inner Mac never needs to sleep!** (Wow, maybe it is Superman).

And if you'd like to try napping or sleeping to let your inner Mac do its thing, we'll have some tips for the best way to make that work in an upcoming chapter. But let's keep to the basics for the moment.

Like **setting a time limit** for your inner MacGyver. We've found it's best if, after writing down your question and telling your inner Mac to work on it, you set a *time limit* for it to do its thing. That can be an hour, half a day, a day or something like that. Too little time and Mac may not have the space it needs to give you really good solutions. Too much time and your inner Mac can get distracted and may drop your particular question or problem. So come back to it in a *set time* to get the ideas you need out before *either of you* move on to something else.

For starting out, we usually suggest at least one to four hours before you come back to ask for your answers. The more you work with your inner Mac this way, you will find the less time it needs to get back to you. I've been doing this for decades now. So it's not uncommon that I can get answers back in as little as 10 or 15 minutes. Which is huge for questions like 'Where did I leave my keys?' Not to mention even bigger and more complex questions. Like 'What is the real theme of the next story I want to write?'

Practice makes perfect. No surprise there. The more you use your inner Mac the faster and better it gets.

But for beginners it's best to give your inner Mac more time, at least until you've really established a good working relationship. So start by walking before you run and then running before you try to fly. And, rest assured, with even a bit of practice, you'll be soaring in no time.

So, **right now,** take a minute or two and write down a few activities that you think might work for you to let your inner Mac do its thing. Again, we're not going anywhere *so just do it now, right here.* Think of tasks you enjoy that let you 'zone out' and lose yourself in the activity.

Got a few listed? *Good!* Because these are the activities you'll be using to practice the secret—at least to get started. You can always add to the list or switch activities if you find one works better than another. Just make sure you give yourself enough to keep busy for the whole time you've allotted. You're in the process of *MacGyvering yourself* here, so shifting gears and innovating to make this work exactly for you is par for the course.

Now that you've written down a problem, and selected a few activities, you're pretty much ready to rip. And in STEP THREE we'll show you the best way to rope in those answers when it's time. But, take it from me, you are mere minutes away from takeoff!

THE SCIENCE: HOW TO INCUBATE?

When you take a break from working on a problem, you are *incubating*,[1] just like an egg is warmed by a hen. A lot is going on inside the egg during incubation, but it's not at all visible on the outside until the chick (or idea) emerges. So, does your mind really incubate ideas?

The power of incubation may be strongest when problems are open-ended, with many possible solutions; however, incubation has been shown to help with tasks such as comparing cars, choosing a roommate or apartment, and detecting when someone is lying.[2] In a recent review of studies on incubation,[3] 85 studies showed that taking a break when working on a creative task reliably leads to better solutions than working straight through.

In one experiment,[4] people were assigned to work individually on an insight problem. When people worked for 35 minutes straight, about 40% solved the problem. If they worked for 13 minutes, took a break for a half hour, and worked again for another 22 minutes, more people solved it (about 60%). But if they worked for 13 minutes, then incubated for three and a half hours before working again for 22 minutes, 85% of the people solved the problem.

In fact, another study claims, "Don't wait to incubate!"[5] Even better performance occurs when you incubate *immediately* after the problem is presented, rather than after working on it consciously for a while. This suggests that subconscious processes can begin working on the problem before any conscious consideration. Just as Lee suggests.

Are some ways of incubating more productive than others? The goal is to reach a state where you can "zone out" and relax into a steady pattern of directed actions. You'll recall that Lee's personal choice is to construct three dimensional models before story writing. He has hit upon a great combination: Studies show that a spatial incubation task benefits a verbal creative task; and vice versa.[6] So, you should do something *different* from the task you are setting loose in your subconscious.[7] That means that you need to customize your approach: Choose an incubation task based on *your* problem and your preferences.

Studies show that incubating by doing an *easy* task (word search puzzles or listening to music) produces better results than performing a more difficult task (anagram puzzles or memory recall task).[8] During incubation, your goal is to keep moving forward on a directed task so that your conscious thoughts are preoccupied; but, it should *not* challenge you. You need an activity that keeps you humming along, is easy to follow, and easy to understand without leading to frustrations and getting stuck. The task that has produced the best results across 92 studies on unconscious thinking is *word search puzzles*.[9] These puzzles can be enjoyable, and since the words to find are listed on the page, they don't require a lot of cognitive resources.

What *not* to do while incubating? Socialize! Listening, watching, reading, any language processing is not well-suited to incubation. Producing language is even more taxing! The reason is that language involves a great deal of unconscious processing within a widely distributed brain network.[10] For example, when you hear the word, "Stop," your mind has to decode the sounds, recognize the word, pull up its meaning, create expectations about what will be said next (syntax and semantics)[11] and interpret its meaning in context. We prioritize determining the meaning of things, so even drawings in comic books can be too taxing. Even when you are not aware of seeing or hearing language, your unconscious processes automatically step in.[12] So it's best to avoid it during an incubation period.

Interestingly, researchers have suggested a method much like Lee's for managing unconscious thinking: "Applying unconscious processes could, for example, entail that people set a goal to find creative solutions for a problem before they are distracted from the problem by doing something different. What people do in the meantime should be chosen carefully."[13]

Like the egg, incubation looks like nothing much is happening; but, inside your mind, your unconscious processes are working through the problem, and leading you towards "hatching" a solution.

HARLEY'S STORY

I'm an art director and graphic designer. And throughout my professional practice I've definitely noticed that creative work happens at odd times during the day – sometimes by my own volition (we creatives are a procrastinating bunch). But also in those four am moments of clarity, or the Creative Director that turns up the next morning with a completely new idea to flesh out. I've often approached creative problems by working hard– and consciously– to a bit of a sticking point, stepping away from it, and then coming back with a fresh take on my work. Although often enough I'd get in a jam and miss the chance for a bit of a break from a project before I had to turn it in. Let's just say I rarely felt like I was getting the best work in on time and was starting to get a bit weary of new projects.

Then I heard about Lee and the MacGyver Secret and got into one of his workshops. I was living in Chicago at the time and would ride my bike to work whenever possible. So I started to write my questions before I left, and would use my bike ride to the office as my activity. And, once I was settled in at my desk, I'd ask for the answers and, sure enough, the things I was after started to appear.

So to come across the MacGyver Secret, and really get familiar with the formal steps of this process, has been a breath of fresh air. And I've been using it for years now. New city, new country even. It still works.

After the initial rush of experiencing what I could achieve with periodic incubations on more and more projects and tasks, the secret has settled down into a fairly regular part of my daily work. It helps me refocus and come up with the goods no matter how close to a deadline, tired or distracted I am. All it takes is the effort of setting some questions, incubating with an activity—which now usually means sitting at my desk and drawing my empty coffee cup badly for twenty minutes– and then taking the time to record whatever my inner Mac has come back to me with. It's definitely made me a lot calmer about creative work and, in a way, I now look *forward* to a really tough problem with lots of moving parts, to give myself a bit of a challenge.

Harley Peddie
Art Director/Graphic Designer

Chapter 8
What is My Inner MacGyver, Really?

So somewhere around now I'll bet there's a question that's been slowly forming at the edges of your mind like, "So you keep talking about my 'inner MacGyver'. But what exactly is that, *really?*"

Great question.

The honest answer is simply this: It's the part of your mind that you are generally **not aware of**. And it goes by many names.

Some call it the 'subconscious' or 'unconscious' mind. As opposed to the 'conscious' mind—or what I call the incessant 'hamster wheel' of the thoughts that run through your mind whenever you're awake. Some call it your 'higher self'. Some researchers call it the 'undermind'. Albert Einstein called it the 'intuitive mind' as opposed to the 'rational mind'. Some even like to think of it as God, whether personal or omnipotent.

So let's try a little thought experiment to see if we can demonstrate the difference between those parts of your mind. Ready?

Right now, count the number of windows in your childhood home. *GO!*

How many did you come up with?

The exact number isn't really important. But, most likely, to find that number you mentally 'walked' through each room, adding up the windows as you went, right?

This is 'conscious' thinking: You are controlling the sequence of thoughts, you are planning what to do next, and making decisions as you go (like, 'do I count the windows in the basement?').

But does this directed thinking also involve thoughts *outside of your awareness?*

Thinking of your childhood home may have raised some emotions that you're not fully aware of, but will likely alter your mood a bit. Perhaps the urge to have some mac'n cheese or some other favorite dish from that time will arise.

And if you now make a list of 'things to do' in the next week, you might be more likely to list 'Call Mom' or one of your siblings. Not because the memory of them was used in counting the windows, but because thinking about 'home' brings them back into your hidden or subconscious mind. With each new thought you have, lots of related ideas can become active in your mind.

This difference between the parts of your mind can also be glimpsed in what is commonly known as a 'Freudian Slip'. To illustrate, Colleen has an embarrassing story of meeting an attractive male scientist at a conference happy hour. She looked at his drink, and noticed a wedding ring. She then meant to ask him, "How long have you been married?" But what came out was, "How long have you been *buried?*" Her subconscious mind was apparently already whirring with interest in the guy, and it was clearly bummed that he was not available. Hence, her conscious mind 'slipped', and the hidden or subconscious thought popped out instead. This guy was 'dead to her.'

While there are a lot of examples like that, the fact is there is **no definitive scientific agreement** about exactly *what*

the hidden or subconscious part of the mind is, what its limits may be, and how exactly it functions. There's lots of impressive research about your brain and, say, which parts control what functions or become active during certain thought patterns or activities. But even the distinction between 'the conscious' and 'the subconscious' is still as much a philosophical question as it is a scientific one.

But you don't have to be a Ph.D. to understand it. You can just think of the MacGyver Secret as a way to establish a *hotline* to that hidden or subconscious part of your mind—your own inner MacGyver. Because we do know that it is constantly aware of and processing *much more* around you than your conscious mind can handle.

As a result, it's in a perfect position to build upon your thoughts and make new, creative connections. *All you need to do is let yourself listen to it* by asking for the answer, and taking the time to *let it speak to you.*

And, when you engage with it as we've laid out, it *will always respond with an answer*. Usually a much better one than your hamster wheel can conjure up, on its best day. Because no matter *what it's called,* it's vastly deeper, faster, and infinitely more creative than your conscious mind.

So, if you'd prefer to think of it as something *other* than your 'inner MacGyver', *feel free*. You can call it George, or Shirley, or my Awesome Awareness or whatever suits your fancy. *It doesn't matter what you call it.* It will work just as well. All you need to do is follow the simple steps offered here and *trust* that the answers will be there when you need them.

After all, how many of us really understand *exactly* how a cell phone works? But we use it all day long. Every day. Because *we know that it works.*

And so will the MacGyver Secret. Even if *no one* yet fully understands exactly why.

THE SCIENCE: THINKING WITHOUT AWARENESS

In the mind, most of what goes on happens unconsciously, outside of our awareness. We only think that all thought is conscious because we (by definition) can't be aware of anything else. Processes taking place outside of conscious awareness[1] are referred to as "nonconscious," "unconscious," "preconscious," and "subconscious" (and now, the term Lee uses, "your inner MacGyver"). Unconscious processes allow us to perform important tasks – like driving a car while talking, navigating a crowded sidewalk, and deciding whether to trust someone – without consciously thinking about them, and even while consciously thinking about something else.[2] Though we value the careful deliberation possible through conscious thought, we do it much less frequently.[3]

To see the unconscious happen, you have to watch carefully; for example, imagine you are out for coffee with a friend. If asked, you would be unable to say what is being said in the conversations going on all around you in the café. But suddenly, you hear someone say your name. Scientists call this the Cocktail Party Effect[4] (apparently, some scientists were once invited to one). This shows that your unconscious processes must have been actively monitoring the sounds around you in order to notice your name and bring it to your attention.

The key point is that fast, unconscious processes – like those taking place during creative incubation – are far more powerful than we realize.[5] When you think consciously about which car to buy, it's hard to consider more than a few things at once; but with unconscious processes, even complex problems with lots of information can be solved.[6] As a result, knowledge becomes better organized, meaning is more deeply understood, and we may arrive at a holistic sense of the best option.[7]

However, we can't easily explain what our unconscious processes do in arriving at a choice, or creating a new idea.[8] In that sense, our unconscious thinking remains a mystery. But increasingly, scientists are identifying how conscious and unconscious processes work together.[9] So Lee's central claim -- that you can "create without awareness" -- is well supported by current science.

EINSTEIN, EDISON AND DALI:
The Genius Trick

Albert Einstein, Thomas Edison and Salvador Dali—all acknowledged geniuses in their respective fields—have all also reported using a surprisingly similar technique when trying to solve a problem. Once having defined the problem for themselves, be it in theoretical physics, mechanical invention, or surreal art, they would sit quietly while holding a small but heavy object: Einstein would use a rock or a marble, Edison a handful of ball bearings, and Dali would hold a heavy key above an overturned plate on the floor. Then they would attempt to forget the problem and drift off into a light sleep, somehow knowing that their deeper, unseen minds (their inner MacGyvers) might hold the solution that their conscious minds had failed to grasp. Once they had moved deep enough into a sleep state, their hands would naturally relax, releasing the object- and causing enough of a noise when it fell- that it would startle them awake. At which moment they would then hope to capture the solution to their question that had emerged from deep inside them before it could slip away.[1]

Each of these stories suggest that, more often than not, the best solutions come—not from their conscious minds—but from the part of their minds *they could not see*. And so were searching for a way to cross that boundary, even if only for a few seconds, to find the answers they sought.

It is unclear where they learned this technique, or if each of them came upon it completely independently.

STEP TWO: SUMMARY

Okay, so before moving on to the final step of the secret, let's do a quick review of how to find the right activity to let your inner MacGyver do its thing:

- Do some enjoyable or necessary activity.
- Keep it physical.
- Keep it unimaginative.
- Keep talking to a minimum.
- Forget TV, reading or really interactive video games.
- Set a time limit.

I know that may sound like a lot to juggle. But, trust me, you'll get the hang of it in no time because, once you find the right activities to keep you busy and let your inner Mac work, the rest of the details will take care of themselves.

The MacGyver Secret

STEP THREE

Chapter 9
Asking Your Inner MacGyver for the Answers

I hope you're ready, because here comes the best part. You've written down your question, told your inner Mac to work on the answer, and kept the hamster wheel in your head busy with some enjoyable or routine activity.

Now it's time to ask your inner MacGyver for the answers. This is even easier than what you've done already. And here are a few tips to make sure you get everything your inner Mac has to offer.

First, **return to your question**. Wherever you wrote your question or problem, in whatever form, **go back to it now** so that it's right in front of you. It really helps to maintain that connection between where you started and what you're about to receive. You can re-read the question if you'd like, or just glance at it to make it fresh in your mind (hopefully, you've really forgotten what it was). Doing this also means you'll be all set to start writing again, so have that pen or pencil handy.

Then, **ask your inner MacGyver for the answer!** Silently, out loud, in song, it doesn't matter how you ask. But *ask*.

Remember, you're trying to create a dialog with your inner Mac, so don't be shy about letting it know you're ready for the answers, and that it's time for it to show you what it's got.

And then, **just start writing!** Anything! *It doesn't matter what you write to get started!* You can write about the activity you just did. You can write the words to The Star Spangled Banner– or any song. You can write how much you love or hate your boss. You can write *anything.* **Just so long as you start writing!**

Because, within no more than a minute of your writing—**The answers will begin to pop into your head as you're writing!** And you will capture them all by *writing them down as they appear.* It's just that simple.

You began this process by *writing down your questions*—and now your inner MacGyver is answering you in the **same way**, *in writing.* This is how you and your inner Mac connect and communicate. Whenever you want it, whenever you need it. You will write what you're looking for help with, and it will **write you back** with the answers. Amazing? Maybe. But true.

And, if the answers you're writing seem to trail off or fade away, or the hamster wheel in your head starts to interfere by focusing on a particular answer, then *just go back to writing anything again,* until the flow returns. Because it will. Just keep that pen going until you're satisfied that you've received all the answers waiting for you.

'But how will I be sure that I've really received all the answers from my inner Mac?' you may be wondering. 'How can I be sure I've written for long enough?' Fair questions. Straight answer: You will **just know**. The same way you *just know* when you get hungry, or thirsty or tired enough to hit the pillow. Your inner Mac will make it clear that its given you all it can, for now.

And if, for any reason, you don't think the answers you've received have fully answered your question or provided enough specifics…**Then ask again!** And tell your inner Mac exactly what you think might be missing, or what you might want it to expand on. In the next chapter we'll give you some tips about *How to Best Evaluate the Answers from Your Inner MacGyver*.

I've also learned there are some questions and problems that **might require your asking more than once**, particularly around personal, emotional, or relationship issues. And we'll cover that too in its own chapter on *Asking The Tough Questions*.

But for now, we'll just focus on the basics for most of your work, technical, or creative questions or problems. All of which should be easily answered if you just keep writing when it's time to receive your answers.

That brings us to the next helpful tip: Just as you gave yourself enough time to really flesh out your questions in detail, **make sure you give yourself enough time to receive your answers**. Remember, you're after really smart and solid answers from your inner MacGyver. Not just *fast* answers.

So, before you ask your inner Mac to deliver the goods, get comfortable, and give yourself the time you'll really need to receive the answers. We've been conditioned by the wonders of technology to expect instant information with the click of a keyboard or the tap on a touch screen. But this information comes from a much deeper and more knowing place—your inner MacGyver. So no point in trying to rush things. *Comprende?*

That's it gang. You now have all you need to make the secret work for you. So, are you ready for this to get real? Because I guarantee your inner MacGyver is!

So **right now,** return to your question, ask your inner MacGyver what it's got for you and just **start writing,** *anything.* And see what happens!

The MacGyver Secret | *Step Three*

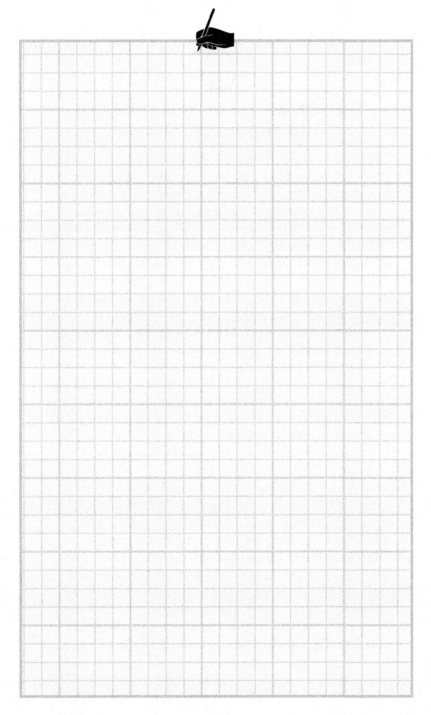

THE SCIENCE: BECOMING CONSCIOUS OF THE UNCONSCIOUS

The process Lee describes for "calling for answers" has limited guidance from science. Historically, understanding unconscious thoughts has been centered in clinical practice. Psychoanalysis[1] was designed by Sigmund Freud for this very problem, where a skilled analyst helps to bring the unconscious into awareness using a variety of strategies, including dream analysis[2] and free association.

A modern method for self-awareness is writing daily in a journal in order to notice the patterns of thoughts on the pages over time.[3] More recently, the practice of mindfulness is studied as a means of becoming aware of unconscious thoughts, and being fully "in the moment."[4] When your mind shifts away from an activity you're doing (meditating or exercising), noticing what you begin to think about may bring you closer to your unconscious thoughts. These approaches may be helpful in improving your ability to bring subconscious thoughts into your awareness.

After years of following this procedure, Lee has ideas pop into mind as complete packages. Other creative types report similar experiences;[5] for example, Paul McCartney awoke one morning with the complete melody for "Yesterday" in mind. But at first, you may experience more haphazard results. Learning to "call for an answer" will take some practice, just like any other skill.[6] The more you practice the sequence of steps with your 'inner Mac', the sooner you will be able to perform the procedure automatically, without having to consciously think about it.

Chapter 10

How to Best Evaluate the Answers From Your Inner MacGyver

Okay, so now you've received your first answers from your inner MacGyver. Maybe what you received totally blew you away. In what we like to call your *MacGyver Moment:* That 'aha' rush of realization or discovery where all the pieces just snap into place, and the answer you feared was beyond you just totally revealed itself like a bolt from the blue. Nothing else quite like that feeling.

Then again, maybe the answers you got back this first time were just...*okay.* Not to worry.

Think back to the first time you tried riding a bicycle. Maybe you hopped on, took off like the wind, and were doing wheelies. More likely it took you a few tries to really get the feel of this new experience: Remembering to peddle, maintaining your balance, steering, and, oh yeah, trying to stop without crashing.

It's no different learning to connect with your inner MacGyver. It may take a few tries and a bit of practice before you're comfortably cruising along, and routinely receiving really solid and innovative answers– in a steady stream of *MacGyver Moments.* As with any new skill or exercise, the more you do it, the easier it gets. And the better the results.

So here are a few tips to help you best evaluate those answers and let you know you're on the right track.

Do any of the answers surprise you?

In the research and workshops we've done with the MacGyver Secret, more than 65% to 75% of the time, people—just like you—get back at least one or more answers that *surprise them*. Saying things like, "Wow. I never would've thought of that!" or "Where did that idea come from?" *Three guesses.*

Because that's one of the real beauties of this secret. Your inner MacGyver can give you answers that your hamster wheel mind *would never come up with!* Real breakthroughs. That for reasons Colleen can better explain, your conscious mind just *isn't designed to give you.*

So if you get back some answers that surprise you, chances are you're right in the groove and you should take those to heart, or really consider following the path they suggest.

'But what if the answers don't really surprise me?' you might ask. 'This was kind of what I was thinking about this problem anyway.' That's fine.

Because it means your inner Mac is *just confirming what you already thought* about the question. And assuring that you have the answer you need. And our research and workshop folks say as much: 'I was sort of thinking that way, but now it feels like that's what my 'gut' is telling me.' Or, 'I knew it had to be something like that, but it feels really clear now that this is the way to go.'

What many people call their 'gut instinct' is just another name for their inner MacGyver. That part of you that somehow senses or 'just knows' what the answer is, even if it might not be fully formed or you can't articulate it. You may not be aware of *how* your inner MacGyver came up with an idea, but now, you can **listen** to it.

Because now, **you can connect straight to it.** And that alone is worth the price of admission.

Does your question just give you back more questions?

This happens fairly often, so don't let it throw you. It means your inner MacGyver is totally on the hunt for the answer you need. But it requires some additional details or info from you to really nail it. So, if you don't immediately know the answers to the questions you got back, then just *ask your inner Mac the questions you received!*

I guarantee that, if you get back questions, they will ***not*** be the same ones you just asked. What's more, it means your inner Mac is really *asking to keep the dialog going* so it can find what you want. So do just that. Keep asking until those surprising—or confirming—answers appear.

'But what if I get back an answer I *don't like?* Or one that kind of scares me?" some of you might ask. While, in my experience, that's rare, it does sometimes happen.

So let me be *ultra clear* about this. **YOUR INNER MACGYVER CAN NEVER HURT YOU, OR MAKE YOU DO ANYTHING YOU DON'T WANT TO DO.** Period. It just can't do that.

On the contrary, it's there to **HELP** you find what you *really* want, about *anything*. Even if it suggests something you may not be fully ready to look at, or suggests a difficult choice.

I'll deal with this in even more detail in the *Asking The Tough Questions* chapter. But if, in that rare instance, you get back an answer that you don't like or upsets you for some reason, then ask yourself ***why?***

'Why does that answer bother me? What about it don't I like?'

Because what your inner MacGyver is trying to tell you is that you might be 'of two minds' about the question at hand, or 'torn' about which way to go. And that perhaps there's some conflict inside you that needs to be resolved.

You are, of course, totally free to ignore it. But why would you want to? If you wrote it down as a question or a problem, then at least *some* part of you—if not *all* of you—is ready to figure this out and move on, or get past it. And you can most definitely use your inner MacGyver to do just that by asking it questions like, 'Why am I struggling with this problem? What really are my best choices about this? What should I do to really resolve this issue?' And so on.

You'll quickly see that, once you get into the habit of asking your questions, then doing your activities of choice, and then asking for the answers, that a smooth and reliable flow of great *MacGyver Moments* will be there. Anytime. Anywhere. Anyhow.

MICHELLE'S STORY

A friend of mine, who is a professor at Harvard, heard Lee give a talk there on the MacGyver Secret. As she knew I was writing a book at the time, she suggested I might want to sign up for one of his workshops. Since I teach, and have a very busy family life, writing has always been something of a challenge for me, so I decided to give it a try.

I was excited to take it even though I wasn't exactly sure what to expect. At first, it seemed like a lot to take in as the process Lee described seemed so different from the way I usually worked.

But, during the class experience, I was fascinated by the idea of asking questions about my work and writing them down, then ignoring them and coming back after an incubation period to call for the answers. It seemed odd, but we tried examples of this in the workshop, and it *worked*. So, although I was a bit reticent to try it on my own creative problems, I did.

The first of these didn't work quite as well as I'd hoped. Perhaps I was still resistant. But each successive time I tried it, the answers came more readily. In fact, sometimes I had lots of answers. Although I had no way at the time to determine if they were great answers or not, they were at least new. In fiction writing particularly, we never know if the answer to a plot problem or character problem is a good one. But you always want to search for something new, something that's not the first idea off the shelf in your mind.

At the end of the first two days of the workshop, I had the most productive writing session that I'd had in a long time, one in which I wrote for 90 minutes in a sustained fervor. This usually happens only about once a year. But it's something that I definitely wanted to repeat.

In the workshop we had long incubation periods that were at least two hours during the day or sometimes overnight. Though we had some sense that these could be shorter over time. I did feel a kind of connection with the work that was new, as though I was able to harness more creativity, more energy in my writing than I'd had before.

I used the method for several months afterward, asking questions at night about life decisions, about creative work, or about teaching problems, all to good effect. It was like having a ready source of advice and it relieved a lot of the anxiety I had about how I was going to address certain problems: How to structure a story or how to cope with some difficulty with my kids.

The hardest part for me was the oddness of the method, which feels sort of like faith. You have to ask the question and admit that you don't have an answer, which is hard. And you have to wait for a while, and then ask again and see what happens. And, even though I got good answers, I'm not sure I ever shook the fear that, at some point, the answers wouldn't be there.

After the workshop ended, I took on a writing project that was emotionally overwhelming. It was a non-fiction book about cancer. I had to devise a nuanced voice for a very difficult subject as well as ways to organize tough information. I even needed strategies to get to speak to my co-authors, who were doctors and were often hard to reach.

I had to do a lot of problem solving on that project, and yet, using the secret, the answers still continued to come. There was a lot of failure in that experience, as well, which is normal in creative endeavors. There were false starts and failed drafts. But I believed that I could find a way, and the answers continued to come because I realized that this is what we all do with writing: We admit that we are stuck and we sort of wait for another answer to appear. I wasn't using the secret as formally as before at that point but now wish I had. Because I think I would have found answers more quickly.

And now, as I'm thinking back on the workshop and the results that I got, I'm feeling that I should go back to it again in that formal way, to see what comes up.

Michelle Seaton
Writer

STEP THREE: SUMMARY

So, to review, here are the bullet points for Asking Your Inner MacGyver for the Answers, and for Best Evaluating those answers:

- **Return to your question.**
- **Ask your inner MacGyver for the answers.**
- **Then just start writing!** *Anything,* **until you're sure you've got it all.**
- **And give yourself enough TIME to receive them.**

And, once you've got those answers down on paper:

- **Does the answer surprise you?** *Good!*
- **Does it seem to confirm what you already knew?** *Good!*
- **If you get back more questions,** *then keep asking!*
- **If the answer seems way off or upsets you, ask** *why?*
- **And, perhaps most important, MY INNER MACGYVER CAN NEVER HURT ME!**

There you have it. Everything you need to make the MacGyver Secret work for *you*. And in the next part, we'll give you even more so that you can easily incorporate the secret into your work and your life. To assure it will always be there for you regardless of the setting, the situation or the nature of your questions.

PART 3

The MacGyver Secret In Action

The MacGyver Secret

Chapter 11

How to Make The MacGyver Secret a Habit

Remember that old joke about a visitor to New York stopping someone on the street to ask, "How do I get to Carnegie Hall?" And the answer they get back is "Practice, practice, practice!" Okay, so maybe not the world's funniest joke.

But the same holds true for the MacGyver Secret. Like any muscle, the more you work it, the more quickly and effectively it responds. And we all know it can be harder to get to the gym or jump on our home elliptical machine *routinely* than we might imagine.

Here then are some tips to make a **habit** of connecting with your inner MacGyver. So you can really reap the amazing benefits it can have on your work—and your life.

Find the right time to write your questions. And do it *daily*.

Some people find the best time to write their questions or problems is just after they get up in the morning. Or before they work out. Or before they commute to work. Others have found it's better to ask when they first land at their desk. Or just before lunch. Or just before they go to bed.

It doesn't matter because your inner Mac will set to work whenever you want. The important thing is to find a time *that works for you*. And then, for at least a week or two, *stick to that time*. That way you can begin to really integrate it into your daily schedule. It will still work even if you have to—or want to—deviate from that pattern. But the people who really get the most from the secret are the ones who have found a way to make it a regular and routine activity.

The same goes for **asking for the answers**. Based on when you write your questions, and whatever activities you do in between, try and ask your inner Mac for the answers around the same time each day by giving it—and yourself—a consistent *Time Limit*. Like after your morning workout. Or right after a quiet lunch. Because, understandably, once you have a fairly regular flow with the secret, the *easier it will be to use it whenever you need to* **outside of that routine.**

As I've explained, I generally used the secret for my writing. But, believe me, it often came in really handy when I was producing or directing a show and a sudden problem arose. Like finding out a location we had planned to shoot was suddenly unavailable. Or an actor we had planned for the day came down with the flu. No matter how well you plan, this stuff happens.

And now I've got a crew of upwards of 100 people—costing thousands of dollars *a minute*—wondering what they're going to do next.

So in a crisis moment, I'd gather all the info I could to fully understand the situation, then tell them I need fifteen minutes, and disappear into a trailer or an empty office and let my inner Mac do its thing. After writing down the question and then doing a crossword puzzle or throwing a tennis ball against the wall for fifteen minutes, I'd ask Mac for an answer. *And I always got one!*

Whether that was giving the missing actor's dialog to another character, or turning a two actor scene into *a phone call—* so I could shoot one side of the conversation until the other actor was available, or quickly concocting a way to *fake* the lost location.

The point is that, because I had *a regular and fluent dialog* with my inner Mac, when my back was really against the wall, it was ready to jump in whenever and wherever I needed it. No need to panic. Just ask Mac.

Now, you will need to experiment a bit to find what best fits the flow of *your* typical day because that can vary widely from person to person. But your inner MacGyver will quickly let you know what feels right and what doesn't. And you can always just **ask it** what time of day would be best for you to write your questions and come back for your answers. Because it will answer that for you as well.

And to help that here's another tip: **Pick a place** for asking and receiving if you can. Perhaps that's a special chair. Or couch. Or at your desk. Or on your bed. But *routinely* writing your questions and receiving your answers in the same *physical place* can only help to reinforce the habit and signal your inner MacGyver that it's time to talk.

The same goes for the activity you choose to let your inner Mac do its thing. If you can try some things, and settle on one or two that really seem to work for you and your inner Mac, *then stick to those* if you can. Again, doing that same activity on a regular basis will let your inner Mac know that it's time for it to be working on your problems.

REVIEW

To review, here are the keys to making the MacGyver Secret a habit:

- **Pick a regular time to write your questions.**
- **Set a regular time limit to ask for and write your answers.**
- **Pick a regular place for asking and receiving your answers.**

- **Pick one or two regular activities to let your inner Mac do its thing.**

That doesn't sound too tough, does it? Didn't think so.

Okay, so **right now**, either **ask your inner Mac** to suggest the best choices for each of the items on that list, or — if you already know what they are — then **write them down now**. And stick to those for at least a week or two to really get the dialog flowing between you and your inner Mac.

Because you have my word, within just a few days of doing that, you'll be having more of those amazing *MacGyver Moments* than you ever thought possible!

The MacGyver Secret

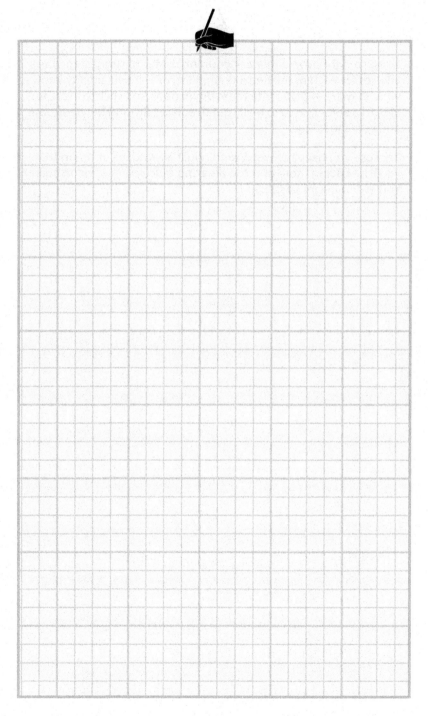

THE SCIENCE: HOW TO BUILD A HABIT

We all know first-hand that changing a habit can be challenging. So what does science say about the best ways to add this new habit to your life?

A key idea is *implementational intentions,* where you plan ahead of time to make the habit happen.[1] For example, if you want a quiet time to commune with your 'inner Mac,' you can think through possible times in the day, and find one that usually provides the needed time and space. Maybe you set the alarm a half-hour earlier, or use the first half of your lunchtime at work.

Try to think of your chosen time and place as a rule: "*If* it's lunchtime at work, *then* I'll 'call for the answers.'" Studies show this really helps people do as they intend; for example, when patients wrote down the plan, "If it is [time] and I am in [place], then I take my pill!" were more likely to actually take their medication on schedule.[2] It's helpful to tie your plan to specific cues that you are likely to notice without extra effort.[3]

Then, you want to repeat the tasks at the same time and place as often as possible; first, get it established, and eventually, you will be able to do it on the go.[4] If you perform a behavior repeatedly in the same context (e.g., taking a walk after dinner), it soon becomes automatic.[5]

SALLY'S STORY

I first met Lee about two years ago when I signed up for a week-long, online course to learn the MacGyver Secret. I'm a former teacher who writes about the challenging intersection between tinkering, mothering, and fear. At the time, I was pregnant and working on my first draft of a book.

I was struggling to finish the last few chapters of my draft before my baby arrived. But, after a few days of trying the secret, the words started coming. Each morning, I would write down what points I wanted to cover in the next chapter and then, as an activity, I would take a long walk. When I returned, I would sit down, ask my "inner MacGyver" for the answers, and start writing. Using the secret, I was able to finish the final chapters of the book much faster than I expected.

Now, as the mother of an amazing and active toddler, I can rarely sit down at my laptop to articulate my thoughts– nor do I want to. So I just use the MacGyver Secret. When an idea for a writing piece comes to mind, I quickly jot it down, sometimes on my calendar. Then I can go back to being fully present with my son, knowing that my idea is not lost. Sometimes the idea incubates for many days, through meandering walks, or another attempt at toilet learning, or around trying to remove blackberry stains. And when I do finally sit down to write, because of the secret, the words are just there for me, and generally flow quickly and efficiently.

Thanks, Lee!

Sally Elley
Writer

Chapter 12

Using Sleep to Let Your Inner MacGyver Do Its Thing

I have a friend who's a heavy hitter in the financial world. No shortage of stress there. Especially when the consequences of any decision could gain or lose *millions*. And, to give himself every advantage on the cusp of a decision, he routinely employs his 'magic couch'. This is nothing more than a comfy sofa in his office that he stretches out on to take a short nap before making a tricky call. And, invariably, when he wakes, his inner MacGyver has told him whether it's time to buy or sell.

And you too can use sleep—either napping or overnight—to let your inner Mac do its thing. Remember, your inner Mac doesn't require sleep. It's there all the time and can easily work on your question or problem whenever the 'hamster wheel' of your thoughts is preoccupied—or shut down completely, like when you're asleep. This is undoubtedly the source of countless breakthroughs and the expression, 'Let me sleep on it.'

Of course, we've found there are some definite challenges to using sleep as the way to get the answers you're after. In my case, for instance, given the pressure I was under to create stories and scripts on a deadline, I needed a way to replicate that process without losing eight or more hours in between. Not to mention that your inner Mac is doing a number of tasks while you're asleep, like dreaming. Which each of us does at least four or five times a night. And retrieving the answers after the transition from sleep to waking can be a tricky process—as you saw from the examples of Einstein, Edison and Dali earlier.

But, if for scheduling or other reasons, sleep is something you'd like to try to let your inner Mac do its thing, here are some tips that might help.

Write down your question *before* you hit the hay, either for a short nap or for the night. Except, this time, in addition to telling your inner MacGyver to work on the problem, you also want to tell it *to remember the answer when you awake!*

All of us have at times awoken from an intense dream only to have it suddenly *evaporate in seconds* as the hamster wheel of our thoughts kicks in. So your dream, that was crystal clear in every detail just a moment ago, is now seemingly *impossible to remember*.

To make sure that doesn't happen to the answers you're counting on from your inner MacGyver, it's important to *tell it to hang onto those for you* in that slippery transition from slumber to waking. Then, just like you would with any activity, **let your question go**, so it's not churning around in your mind before you try to sleep. Your inner Mac will get to it even while your sleeping. Truly.

It also helps to **keep your notebook and writing stuff nearby**—like on your nightstand—so you can grab it as soon as you wake up. That way your question is right there and you can **start writing to receive the answers** as quickly and easily as possible after your eyes pop open. And it's okay to start by writing what you can remember from your dreams, or whatever first comes to mind, in order to draw out the answers from your inner Mac through your pen.

This method will also definitely require a bit of regular practice. So don't lose heart or give up if it takes a few nights—or multiple naps—with these added instructions to your inner Mac—before the answers to your questions begin to flow smoothly.

You've been falling asleep and waking up your entire life. And so the parts of your mind have developed a consistent pattern for that process. Now that you're trying to alter or refine that process by asking for something specific from it, it may take several tries before you've successfully changed or 're-trained' that pattern. But you can. And your inner MacGyver is only too willing to work with you on that.

REVIEW

To review then, here are the key tips for using sleep to let your inner Mac do its thing.

- Write your question just before you go to sleep.

- And make sure to tell your inner MacGyver to remember the answers when you wake up in addition to working on your problem.

- Then let the question go. *(Don't think about it any more!)*

- Keep your writing stuff close at hand so you can start writing as soon after you awake as possible.

- Don't sweat or fret if it takes a few tries to get your answers this way.

Now, tell me it isn't awesome to think that you can be using your inner MacGyver to solve your problems *while you sleep*. Every night! Talk about making the most of your downtime, huh?

THE SCIENCE: USING SLEEP TO INCUBATE

Of all incubation activities – distracting from the task, enjoyable, as long as needed – sleep seems ideal. Can sleep help you incubate on problems?

Incubation during sleep does show benefits for complex problems. In one study, people learned a sequence of moves in a game with a hidden shortcut; 60% of those who slept for 8 hours after the training discovered the insight, while only 24% of those with no sleep (during daytime or else at night) did.[1] In another study on finding remote associations,[2] a sleep group solved more problems than no-sleep groups, but only with difficult rather than easy items. The scientists conclude, "**Sleep on it, but only if it is difficult.**"

And not just any sleep: About a quarter of the time asleep, people enter into rapid eye movement (REM) sleep. Your eyes move rapidly and at random, your brain's activities are similar to when awake, and your most vivid dreams take place. REM sleep includes hallucinatory and bizarre content.[3] We usually miss out on REM sleep if we don't sleep restfully or long enough.

But REM sleep appears to be the active ingredient needed as an incubator for creativity. One study compared solution rates for attempted (and unsolved) remote associate problems. Those who napped with REM sleep improved by almost 40%, but the non-REM sleep and "no nap" groups showed no benefit.[4] REM sleep appears to integrate unassociated information in memory, so may lead to increased creativity.

The challenge in using sleep to incubate is that you need to call for the answers right away, before they are lost as you become fully awake.

JACK NICKLAUS, ELIAS HOWE & NIELS BOHR:
Finding Solutions in a Dream

JACK NICKLAUS' STORY:
Regaining his Golf Swing in a Dream

In 1964, legendary golfer Jack Nicklaus uncharacteristically fell out of form. Nicklaus was not winning in tournaments because he was hitting scores in the high seventies. Then, all of a sudden, his prowess returned. Nicklaus explained the sudden improvement like this:

> "Wednesday night I had a dream and it was about my golf swing. I was hitting them pretty good in the dream and all at once I realized I wasn't holding the club the way I've actually been holding it lately. I've been having trouble collapsing my right arm taking the club head away from the ball, but I was doing it perfectly in my sleep. So when I came to the course the next morning I tried it the way I did in my dream and it worked. I shot a sixty-eight yesterday and a sixty-five today."[1]

ELIAS HOWE'S STORY:
Sewing Machine Breakthrough in a Dream

Inventor Elias Howe had a great idea about a machine that could be used for sewing. His concept consisted of a needle attached to a machine that would go through material. Initially, Howe tried using a needle with a sharp point at both ends and an eye in the middle, but it failed.

Howe couldn't quite link the entire project into a working model until one night he had a dream. In the dream Howe saw himself being taken prisoner by a group of savages. He watched as these men danced around him with spears and noticed that each native's spear *had a hole near the tip.*

Upon awaking, Howe instantly recognized the dream had solved the problem to his proposed sewing machine. That is, via placing a hole at the tip of the needle, the thread could be caught after piercing the material. He instantly set about changing the design of his machine and after making the necessary alterations, discovered it worked![2]

NIELS BOHR'S STORY:
Discovering the Structure of the Atom in a Dream

The physicist, Niels Bohr, got his doctorate in 1911 and gained notoriety for deciphering complex problems in the world of physics that had left his colleagues stumped. In time, he set upon understanding the structure of the atom, but none of his configurations would fit.

One night he went to sleep and began dreaming about atoms. He saw the nucleus of the atom, with electrons spinning around it, much as planets spin around their sun.

Immediately on awakening, Bohr felt the vision was accurate. But as a scientist he knew the importance of validating his idea before announcing it to the world. He returned to his lab and searched for evidence to support his theory. It held true - and Bohr's vision of atomic structure turned out to be one of the greatest breakthroughs of his day. Bohr was later awarded a Nobel Prize for Physics as a result of this leap in creative thinking *while asleep.*[3]

Chapter 13

Asking Your Inner MacGyver Tough Questions

While it may seem at first glance that all the questions you might choose to ask your inner MacGyver are the same, the fact is they're not.

For instance, there's a big difference between a question like, 'What's the best outline for this report I need to write?' and, 'Is it time to split up with my partner?' The first is a fairly simple creative question about a layout. The second is an intensely personal question that most likely is packed with a lot of emotional baggage.

As are questions like, 'Why am I so unhappy all the time?' or 'Why do I have so much trouble making friends?' or, 'Why can't I find and maintain a solid romantic relationship?' or, 'Why can't I get past my addictive tendencies for things I know are bad for me?' We've all asked ourselves these kinds of questions. Usually more than once. Because these are tough questions.

Fortunately, your inner MacGyver can help you find the answers to these questions too. For real.

So here are some tips that might help if that's the way you want to go.

Tough questions often require that you ask them of your inner Mac more than once.

These types of questions generally involve some painful memories or experiences. And one of the essential jobs of your inner MacGyver—or whatever you choose to call it—is *to shield your conscious mind from that pain.* In part, because if you had to face that emotional pain all the time, it would be extremely difficult to function.

Since your inner Mac is helping to protect you from that pain, to really assure it that *you're now ready to face those questions,* you may have to ask them more than once. Usually no more than two or three times at the most. At which point your inner Mac will know you're determined to deal with these difficult problems and will offer you the answers you need.

For example, one of the people in a workshop was considerably overweight and he really wanted to know why. The initial answers he got back from his inner Mac were vague and unsatisfactory. But, determined to get a helpful answer, he agreed to continue to ask the question over the course of the next week or so. And, shortly, the answers began to emerge.

He realized he was afraid that he wasn't being effective or taken seriously at his job. And eating was a way to hide that fear behind the pleasure of food. And that his growing size literally created a 'wall' around him to 'distance' himself from his co-workers and retreat from the fear. He was, in effect, trying to hide in his own skin. And now knew it.

So he tried using his inner Mac to improve his performance at work and make more of an effort to connect with his colleagues. And once he sensed that his 'status' and position at his job had taken a turn for the better, food became less of a need or a solution, and he could begin to forge a plan to tackle his weight problem.

Needless to say, answers like those may not be easy to consider or act on. You are, after all, asking yourself some tough questions. But I can assure you, the answers you receive will be *genuine, honest and truly helpful,* and can only build a stronger bridge between your waking or conscious mind and your deeper, inner self. Because both parts of you will now be starting to *align themselves,* rather than continuing to be at odds.

In addition to possibly having to ask a tough question more than once, **it may also take more time to get back the answers** you're looking for. That could be a few days, or a week, or more.

Another example: One of the participants in our online workshops had a difficult relationship issue. After suggesting she ask the same question multiple times, and telling her to be patient with her inner Mac, I got an email a week after the workshop ended. She explained that *two weeks from the day* she had first asked her question, she was walking down the street when the answer came to her 'like a bolt from above'. And she realized that she now knew *exactly what she needed to do* with regard to her relationship. Her email closed with an exuberant, 'THANK YOU!'

No thanks needed. *She* had the courage to ask the tough question as did the person who was stressed about his weight. And their inner MacGyvers had given them the answers they were looking for. As it will for you.

I usually advise that you get a good dialog going with your inner Mac—posing questions and receiving answers daily– for at least a week or two before asking it a tough question.

But, if it's really a problem that's been hanging on you, then you can go for it whenever you feel the need.

REVIEW

So, to review, here are the key things to remember if you decide to ask your inner MacGyver an emotionally charged or personal and difficult question.

- Be prepared to ask your inner Mac the question more than once.

- Be patient with it while waiting for the answer.

- Know that, sooner or later, *you will get the answer you need!*

And, if you're thinking there might be some tough questions you'd like to ask your inner Mac, rather than diving in and asking them right now, you can **just list one or more of them now.** Or just list some problems you'd like help with *at some point.* And then come back, say, in a few weeks after you've had some practice connecting with your inner MacGyver, to 'officially' ask for help then.

The fact is, just by listing something tough, you've already cued your inner Mac to gently start the process. But, at the risk of repeating myself, it's better to work up to those slowly then plunge in with no practice at all.

ALLEN'S STORY

I'm a foreign exchange student from Zimbabwe and have recently earned my bachelor's degree in the United States. I first learned about the MacGyver Secret at a conference where Lee gave a talk about it. As part of his introduction, he asked everyone in the audience to write down a question we had been trying to answer for ourselves, about anything.

Having just graduated, I was at a crossroads and was struggling to decide what to do and where to go with my life: Whether to go on for a graduate degree or become an entrepreneur. Lee said to be as specific as we could in describing our question. And, as I began to write it, my question somehow got wider "How do I live a life of purpose?" And hoped I would get an answer to help me move on.

My first incubation activities were driving and sleeping. I drove home from the conference that day, spending an hour on the road. I relayed my experience at the event to my girl-friend then proceeded to bed. In the morning, I decided to put the MacGyver Secret to the test.

I sat down to write the answer and began by putting my pen down and simply writing "So". *And the rest just flowed out of me.* I was surprised by the depth and detail of the answer and how accurate it was in relation to past events in my life and my attitudes, many of which had been hidden inside me.

The answer also had a question, so I asked that question separately, fleshing it out and making it more specific. Then I proceeded to do some yard work and cleaning the house. Later that afternoon I sat down to call for the answer and again was astonished at the detail and the depth of thought involved.

The first two weeks of using the MacGyver Secret led to my apartment being spotless. I got into the habit of watering plants as well and felt more peaceful as I put the secret to use on all that was bothering me.

I have been using it for five months now, at least once every two weeks if not more often than that. I use it mostly for personal questions, like psychological or relationship issues. My preferred incubation activity now is sleep. Sometimes I nap or sleep over night. Other times I sleep on it for a few days before coming back to call for the answers and write them down. I have received so many beautiful, philosophical and interesting answers. And I've found myself less stressed out about the situations I'm facing in my personal life.

I have benefited so much from this wonderful tool—*that works when I'm sleeping.* It's almost like outsourcing a question to a foreign country were people are awake and working on your question when your sleeping. Then you wake up and 'voila!!!' there is my solution and my path. I am excited now to try using it as I move into my professional life.

Allen K. Matsika
Exchange Student

Chapter 14

Using The MacGyver Secret in the Workplace

Many of us work in offices from 9 to 5—or longer—where 'productivity' is generally defined by the number of hours we remain glued to our desks and computers. And in these, or other restrictive work settings, it might appear difficult to really employ the MacGyver Secret. Fear not.

Because there are a number of simple tips and strategies to successfully tap into your inner Mac in these situations so neither your co-workers—nor your boss– *will ever know you're doing it!* Think of it as the Mac Secret in 'stealth' mode.

The first thing to consider is **finding the right time to write your questions**. This might be before you leave home for your morning commute, when you first arrive at your desk, or perhaps before lunch.

The real trick here is being able to easily transition into the **right activity** to let your inner Mac do its thing. We call one such activity *Going On Tour*.

By that we only mean leaving your desk to visit the office kitchen, or the bathroom, or taking a stroll around the office to make coffee, turn in recyclables, or make copies. If done one after the other, all these excursions could easily add up to 30 to 60 minutes. And, with even a little practice, that should be enough for you to return to your desk, ask for the answers, and get the valuable responses you're after.

Of course, if such a *'tour'* proves too distracting for you—or draws the unwanted attention of anyone (like your boss)—then there's another trick we call *Swabbing The Decks*. This means doing any number of mundane office activities like cleaning off your desktop, filing, routine data entry or cleaning the office kitchen—which will no doubt endear you to your workmates if not possibly earn some points with the honchos.

The fact is most office jobs entail more than their share of important but mind-numbing tasks. Most of which we tend to avoid because of how dull and boring they are. But these turn out to be *perfect activities* to let your inner MacGyver do its thing.

So even if you should run out of your own stuff like this, you can always volunteer to take on those of a co-worker. Or offer to do them in exchange for something you might want from them. You can volunteer to do the 'run' to pick up lunch, or pick up a visitor at the airport, or drop off a package at another building. This is not only a great way to win friends and influence people, none of whom realize that, by doing their 'scut-work', you are secretly being *incredibly productive* at coming up with some great ideas to blow everyone away at the next meeting. A total Mac-hack!

Still need more time? Then what about using your lunch break as the activity to let your inner Mac do its thing? So, instead of using that time to chew the fat with your buds, simply go out or eat in by *yourself*. And do a crossword puzzle, wordsearch, Sudoku or some other similar activity while you eat. Be sure to add on a walk to and from the nearest (or farthest) deli to help do the trick. With a bit of practice, you should have no trouble getting the goods from your inner Mac after your feeding break.

Then again, if you think your boss might be open to some new ideas about how to improve 'office productivity', you can always try slipping them a copy of this book. And say something like, 'You know, I've been playing around with this at home—and it *really works!* You think it might be something that would help kick things up a notch around the office?' That way, if your boss sparks to it, then it becomes *their idea* to share with everyone. With any luck, you'll soon be out of stealth mode, sharing the fine points of the Mac Secret with all of your colleagues, and subtly building the runway for your next promotion. Just a thought.

REVIEW

To review then, here are the highlights to making the secret work for you at the office.

- Find the right time to write your questions.

- Then go on tour, swab the decks, or use your lunch break as the activity to let your inner Mac do its thing.

- Then call for the answers.

And if you still need help with this, then **just ask your inner Mac** how to best make it happen in your particular work place. Because, unless you're a deep-sea welder or a jet fighter pilot, the odds are you can find a way to make the secret work, even on the job.

So, if you're thinking you'd like to try the MacGyver Secret at the office, **just write down now** which of those tactics you think might work best for you. Or *ask your inner Mac* for some ideas that would better fit your situation. And you should be all set to give it a whack the next time you suit up for the office.

Chapter 15

Using The MacGyver Secret With a Team

Working in a small, focused team has been steadily gaining popularity in a host of fields, particular among tech and other entrepreneurial companies. And there is a growing body of research that supports the effectiveness of such an approach. Though there's still some debate about why some teams turn out to be successful while others don't seem to 'gel' at all.

Either way, if you find yourself part of such a team, here are some tips to use the MacGyver Secret, and make the most of that approach both for yourself—and your team.

Obviously, if you don't think your team will be open to diving into their inner MacGyvers *together,* you can always just do it yourself. Once you have the topic or goal to be 'brainstormed' at your next gathering—and the time you're planning to meet– then you can simply do a 'cycle' or two with your inner Mac (question—activity—answer) sometime before then. You'll walk into the meeting armed with some great ideas to get the ball rolling for the group.

Ideally, you can offer to *share the secret* with your team as a new way for *everyone* to approach the task at hand. And, perhaps, change up the old work patterns to generate some amazing breakthroughs *as a team.* So here are some tips to pulling that off.

Once you've convened together, and agreed on the topic or problem to be tackled, **then each of the team members should write that down in their own words—*separately*.** So each team member has their own notebook or paper and defines the problem *individually*.

Why? Sometimes even subtle differences in the way people describe a problem can lead to radically different potential solutions. And you don't want to inhibit or 'short circuit' that possibility by all describing the problem in the exact same way—or words. So you don't want to copy the problem that one person or the team leader might write on a whiteboard, etc. Let each team member define the challenge in *their own way*.

Then, having each described the question or goal in detail, in their own terms, **don't discuss it—*at all!*** Just take a moment for everyone to silently tell their inner Mac to work on it, agree on a time limit and then either—

Find an activity for the group to do together to let everyone's inner Mac do its thing, or **let everyone go off and do an individual activity.**

Some group activities could be playing a dice or card game (but keep talk to a minimum), going for a ***silent*** walk or stair climb together around the office building, campus or neighborhood, or each work on improving skills like, say, sketching.

Group sports are a great task, such as playing a quick game of basketball or volleyball. Ideally, as at firms like Google, your office provides a Ping Pong table where you can play a round-robin game of Ping Pong (where you substitute for each other after every few points, etc.). You get the idea. You're looking for some group activity that has some physical aspect to it and will keep all of your hamster wheels occupied so *you're all unlikely to think about the question at hand.*

Or, if that's not feasible or desirable for the group, then *everyone can simply do their own activity,* either separately or in the same place.

That's how we do it in our workshops, with each person finding their own space at a table, or on a couch, or the floor. They each choose their own activity, such as building a simple Lego set (always a hit with the guys for some reason), doing a crossword or Sudoku puzzle, doing a small but tricky jigsaw puzzle, creating something with paper from a set of origami instructions, or following the instructions in a 'Learn-To-Draw' book, with dinosaurs, people, flowers, whatever.

Then, once the time limit is up, you all reconvene, silently ask your inner Macs for the answers, **and then each write your answers down– *individually,*** just as you did with the problem, taking as much time as needed for everyone to start writing until the solutions begin to flow for each of you.

After that, each of you can go to the whiteboard and write your ideas, or you can go around the room and read each of your answers to the group. Better still, copy each of your answers onto a separate sheet, pool them all in a stack, and shuffle them. Then take turns pulling a sheet from the pool and reading it aloud. That way, every idea is heard, but also *anonymous.*

Then—and only then—should you begin to discuss them as a group.

The reason for this is simple: What frequently limits the success of a team are the *interpersonal dynamics* of the group. Where some members are stronger, more forceful voices and others more shy or reticent to share their thoughts. This can easily skew the discussion, and sometimes push the team in a direction that is determined more by the personalities within the team than the best approaches to the problem at hand.

So by generating the answers *individually* **before** sharing them with the group, you can 'by-pass' those differences in personality and really get *all the best ideas out on the table* before deciding as a group which might be the right ones to pursue.

Doing it this way, you get to *multiply or compound* the value from all your inner MacGyvers rather than succumbing to just the thoughts of the loudest members of the team.

REVIEW

To review then, here are the tips for making the most of the MacGyver Secret with a team.

- After a brief discussion of the problem or goal each member of the team should write/define that in their own words, before each asking your inner Macs to do their thing.

- Then, don't discuss the question — at all.

- Move on to an activity, either for the group as a whole or individually.

- Then, after coming back together after the time limit, each member of the group should ask for the answers and write them down individually.

- Then — and only then — you can share your answers, ideally, anonymously.

I can all but *guarantee* you that the discussion following this approach will be remarkably free-flowing if not mind-blowing. Because many of the answers that emerge will be fresh and surprising—*even to those who came up with them!* Which can only help the group see the initial problem in innovative and unexpected ways. Like Mac to the power of five!—Or however many there are on your team.

THE SCIENCE:
IS GROUP 'BRAINSTORMING' BETTER?

People assume the power of the group will far surpass the individual. Certainly, people enjoy working with a group, and building on others' ideas. But how do we know that groups are better?

A great deal of research has established that the individuals working alone far exceed the results of the group, both in quantity and quality. For example, suppose eight individuals are asked to do a task (e.g., come up with a new brand name for spaghetti), and when they are done, their work is compiled. Then, their results are compared to another group of eight who worked together for the same amount of time. The manpower is the same, but what about the idea power of individuals compared to groups?

Studies show that working individually is *more* efficient than collaborating in groups; in fact, *group process loss* holds up across settings and tasks.[1] When working with the group, people worry about how their ideas will be evaluated, have to wait their turn to talk, and every group seems to include the dreaded "free rider" who doesn't contribute. [2]

We have an "illusion of group productivity."[3] The recommendations from science for making the most of creative minds is to start with individual work first to get the most and best ideas out, and then share so the group can build from them. Just as Lee suggests!

The MacGyver Secret

Chapter 16
Feeding Your Inner MacGyver

Without question, your inner MacGyver is an almost bottomless well of amazing stuff. All the answers you need may well be *within you;* you just need to *ask for them!*

This is because your mind contains—and retains—*all the accumulated experiences of your entire lifetime.* Your inner MacGyver (or subconscious) records and processes *all of it while it's happening;* whether during focused learning or thought, while you witness it happening to someone else, and even if it's happening around you and *you are not consciously aware of it at all!*

Believe it or not, *it's all in there.* Everything. So it's no wonder that the sum of your experiences can provide connections among ideas that, upon your asking, may often surprise you.

And, if you'd like to kick your inner MacGyver up another level—to, say, *turbo Mac*—you can do that too…by 'feeding' it. Meaning that you offer your inner Mac some key elements for the challenge you're about to hand it. (In Colleen's work, this is called 'opportunism').

For example, I was once confronted with a particularly difficult writing problem on a movie. Typically, in the world of independent features, one writes a script and then goes looking for the money to make it. But, in this instance, it was just the reverse. There was a group who had the money to make a movie but had no script—or even a story!

All they knew was that they wanted a strong and moving character-driven drama. And it fell to me to figure out exactly what that was. Talk about starting from scratch. Seriously, I had **nothing**.

So before turning to my inner Mac to construct a story, I just started looking for any interesting story elements that caught my eye. If the goal was to write a character driven drama, then I knew I at least needed an interesting character who found him or herself in some unusual or difficult situation—for the drama part.

The first thing that sparked my interest was a newspaper article about Appalachian slums in Ohio cities. It seemed when the coal mines of Appalachia stopped producing, many of those people migrated north looking for work and found themselves ghettoized, if not discriminated against, in a number of cities in Ohio. I had no clue what to do with the article but it was news to me so I threw it to my inner Mac.

Then I stumbled on a blurb in a magazine about a number of states who, to save money, had the inmates in their prisons answering calls from tourists who were considering visiting their state. Come again? So, if, say, I was calling to find out the best places to check out in Kansas, I might find myself talking to Elwood who was actually doing 3-5 for aggravated assault or armed robbery. Again, I had no idea what to do with that except it was too crazy not to throw into the Mac hopper.

Finally, when I was driving home from the office one evening, I heard a story on NPR about some people who owned an inn in Maine they had been trying to sell but couldn't find any buyers. So they decided to run an essay contest: If you sent in $100 and an essay about why you wanted to own their inn—and they got at least 2500 such essays—then they would give the inn to whoever wrote the best essay. Well, much to their surprise, they got 7000 submissions! Crazy, huh?

The MacGyver Secret

So, having 'fed' those and a number of other random pieces to my inner Mac, I then turned it loose and told it to come up with a great story for this movie. Then I went to work on a paper model of the Statue of Liberty—or maybe it was the Great Sphinx of Giza. The stories I remember, the models not so much.

And what I got back was this: What if there was a girl from an Appalachian slum in Ohio who found herself answering tourist calls from a prison in Maine? And when she gets out of prison, she goes to a small town in Maine where she gets a job working at a local diner, and helps the aging owner of the diner 'sell' it by running an essay contest?

A good start. Needless to say, I went back to my inner Mac any number of times to fill in the connections between all these pieces: Why was she in prison? Why did the diner owner want to sell it? How did this create a conflict? Who were the other characters? And so on. But now, I had the makings of a potentially really cool story.

The point of all that being, when you know you've got a big project or problem, you can 'feed' your inner Mac with as many ideas or elements that you think might prove of value to the solution. No matter how 'outside the box' or tangential they may seem. The more the merrier. After all, that's Mac's specialty, no? Taking random things and combining them to come up with a totally unexpected and innovative solution.

So set out to **expose yourself to unfamiliar ideas.** You can do this by traveling to new places, trying new activities (never done karaoke?), or volunteering at a new community group.

You can even do this at home by switching up what you read, or visiting web sites just to see what's out there (try 'Stumbleupon.com', which will send you to another random site with each click). Novelty is the ingredient that **makes you slow down and pay attention** to how things work around you. That, in turn, leads to interesting new thoughts and ideas.

Then **notice what you notice**. What do you find interesting or of note? If your best friend was sitting next to you, what would you comment upon to them? What matters is what you notice—there is no 'right answer' here. You simply pay attention to the things that catch your attention.

Each idea you think about becomes more food for your inner Mac. And, believe me, *it will all get in there* for Mac to use. Like that spaghetti sauce commercial, "Your kids won't know it, but all the vegetables are in there!" Same with your inner Mac. Anything you 'feed' it is *in there*.

Then, do the Mac Secret. Same as always, pick your question, write it down, give your inner Mac some space, and then call for the answers to see what comes up.

Even if what you fed it doesn't show up right away in the responses you get back, *it's still in there*. And consistent attention to novel things you care about will ensure that your inner Mac stays alert to good ideas.

*So feed it—all the time—*on a steady diet of thoughts or ideas that spark your interest.

As it was in my case, this can be especially helpful if you're really starting with a 'blank slate' and may be uncertain exactly where to begin on the problem.

And, in case you were wondering, that movie, which I also directed, was *THE SPITFIRE GRILL*. It went on to win the Audience Award at the Sundance Film Festival, broke the record at the time for the most money ever paid for a movie from Sundance, and was then turned into a *musical,* of all things, that has had over 600 productions so far—and still counting. Go figure.

So if you don't think your inner Mac can really help you hit one out of the park, **think again!**

THE SCIENCE: THE PREPARED MIND

For creativity to happen, the "raw materials" have to be available in memory for your unconscious processes to explore.[1] But not just any random combination: As Louis Pasteur put it, "Chance favors the prepared mind."[2] What he presumably meant by this is that prepared minds can and do take advantage of lucky encounters with relevant external events.

Lee's examples in this chapter illustrate this process of opportunistic assimilation[3]: finding an interesting story, thinking about its meaning, moving on, and then later, asking a question while creating that leads to a connection to the story in memory. At the time you see a new idea, you need to consciously process its meaning, elaborating on the information to help connect it to your goals. Why is it of interest? In laboratory studies, this predictive encoding at the time results in automatic access later when those circumstances come up.[4]

In Lee's example, he read about prisoners working as state travel representatives; likely, he was puzzled. What would it be like to "sell" a place you can't visit? If you were the prisoner, how would you feel about the place you're advertising? This conflict made it a great candidate for later creative use with a story character. By anticipating when the new idea might be useful in the future, you increase the potential to connect with this idea when it's needed.[3]

Making a regular practice of seeking out new information (reading news stories, online posts, seeing artwork, etc.), helps you establish a knowledge base of interesting ideas in memory.[1] Then, with some conscious work to uncover what makes them interesting, you can prepare your mind to connect to it later in just the right circumstances.

The MacGyver Secret

Chapter 17
Life With Your Inner MacGyver

I've been teaching the MacGyver Secret for several years now—to hundreds of people—in all walks of life and professions. In part, I wanted to gain valuable feedback for the preparation of this book, so that you would have not only the benefit of my experience with it, but theirs as well.

As you might imagine, a number of interesting questions have come up in those workshops and seminars that are worth sharing as you embark on your journey with this remarkable process.

For instance, a number of people have asked what the *real* goal or purpose is of the different incubation activities we suggest you try. Simply put, the goal of whatever activity you choose is to 'zone out' and get lost in the moment to moment execution of what you're doing. So that *you're not thinking about the problem* you've given to your inner Mac. Instead you are *focused on what comes next in your activity*, whether that's the next step in a set of instructions you're following, or the route of your jog, or the notes of the music you're practicing on an instrument, or the next ingredient you'll need to add to the dish you're cooking.

And, along those lines, people who engage in various forms of *meditation* on a regular basis were curious if that could be used as their incubation activity?

The answer is, *"yes!"* The key to most forms of meditation is to remain conscious but *without thoughts*. (What I call the "hamster wheel" in your head). So, in effect, all of the activities we've suggested are designed *to do the same thing:* to remain conscious without focusing on the problem you've turned over to your inner MacGyver. So, if you happen to meditate—or are thinking of taking that up for all of the benefits it offers—then you can also use it as *one of your activities* to let your inner Mac go to work and do its thing.

Also, since all of the workshops are at least a day—if not several days—they all involve breaks for meals, like lunch. And one of the questions that often emerges is, if using the lunch break as an incubation activity—which is something we've suggested earlier in the book—is it okay to chat with their friends and colleagues during that break? Sorry, not really. Since conversation or social interactions require a lot of your inner Mac's attention, lunch is only an effective activity if you are by *yourself and not talking or listening to others*. You can do a puzzle during lunch or play a simple game on your phone, like Candy Crush, but it's really best to avoid reading or talking. So, if you want to talk with your friends during lunch, then find another incubation activity to let your inner Mac work for you later. Because all that interaction won't allow it to solve the question or problem you've asked it to tackle for you.

Then there's almost always the question about 'writer's block'. Lots of writers and other creative types are drawn to the secret because, like me, they are burdened with having to constantly create new material. And, they want to know if the secret can really help overcome those frustrating bouts of writer's block. To which I always respond with *"Absolutely!"* It is, in fact, the *ideal antidote* for writer's block. Because the 'block' you're experiencing is that your hamster wheel—or conscious mind—is so caught up by anxiety, frustration, fear or some other emotion that it's 'dammed up' the flow from your creative source—or inner MacGyver. By using the secret to 'bypass' your conscious mind entirely, you can access that creative source *directly* and get the 'flow' of words and ideas going again.

Then, usually as the workshop is winding down, someone quietly pipes in with a question about alcohol and drugs, like marijuana: Can any of those be used with the MacGyver Secret? Well, the fact is alcohol and drugs have been used for centuries by creatives, like writers, artists and inventors to soften or 'bypass' the restrictive and critical nature of their conscious minds, and allow the work to flow out from their inner self or subconscious. Of course, much of the beauty of the MacGyver Secret is that it can produce that same flow *without the need for drugs or alcohol.*

That said, there is research to suggest that, in moderate amounts, such inebriants can sometimes facilitate the creative process.[1] So, if you're of age, and these substances are legal where you live, here's my advice if you want to experiment with adding those to the mix. First, practice the secret clean for at least several weeks to develop that dialog and flow with your inner MacGyver. Once you're comfortable enough to know the answers will come to you regardless, then, *just before you ask for the answers,* you can have a drink or some cannabis to see if that increases the flow or the scope of the answers from your inner Mac. For sure though, too much of any drug will simply hammer both aspects of your mind to the point where nothing of value can emerge, nor will you be likely to retain whatever insights might appear. So, to make the lawyers happy, *I am not in any way recommending that you try this!* Nor is it in any way necessary to use drugs or alcohol to make the secret work. But, should you decide to, I would suggest following these guidelines.

And, once the ice has been broken with the subject of alcohol and drugs, invariably the very next question is about, you guessed it, *sex! Can sex work as an incubation activity?* The room goes silent as every eye is suddenly riveted on me, awaiting the answer. And I let that breathless pause hang for a beat, just for the sheer drama of it, before I reply with a reassuring smile… "Yes, *indeed,* sex can be an effective incubation activity." Why? Because it fits all the criteria: It's physical, enjoyable, and keeps your conscious mind focused on something other than the questions you've posed to your inner Mac. So, assuming of course that it's *completely consensual,* you can try sex as an incubation activity. Though, so as not to freak out your partner, it's better if you don't yell 'Eureka! I've got the answer!' while in the midst of things. Best to wait until you're well into the 'afterglow' before reaching for your notebook to start writing the answers. Alas, watching pornography, like watching any video, *will not work* so, if you don't have a willing partner, I'd suggest finding another activity you can do solo.

Finally, as the wave of nervous joy and relief about the sex question ripples around the room, there's one last question that has come up in every talk and workshop I've ever done about the secret. Which is this, "Has it really made *that much* of a difference in your life, Lee? Can something as simple as you make this sound really have *that big* of an impact?"

And, to answer that questions, here's what I say, and the last story I will leave with you…

Let's face it, sooner or later we all have problems. Maybe with work, or relationships, or money or *something.* Nobody's life is perfect, even if it may seem like it is. Trust me, if you're human, you have problems. And having access to your inner Mac to help with those is a great thing.

But that's only the half of it. Because, once you've really opened a solid dialog with your inner Mac, and learned to trust what it can offer you, *that's when its amazing bonus benefits begin to reveal themselves.*

Now I tend to be a pretty practical, hands-on, kind of person. It's not that I have anything against the touchy-feely, spiritual, mystical or magical way of looking at things. If that works for you, by all means, go for it. I'm just more comfortable with the nuts and bolts approach: If something's broken or not working, then just find the right tools and fix it, whether it's a leaking faucet or a lousy script. Like that famous acronym KISS: Keep it simple, stupid. That's me.

So I'll admit it took me a while to wrap my head around fully believing in my inner Mac because *it seems like magic.* Ask it a question and get back a great answer? Repeatedly? Without really knowing exactly how and why? It just seemed too good to be true, you know?

But *seeing is believing.* And, after all the scripts, and success, and rewards, and *awards,* there was no getting around the simple fact *that it works!*

And perhaps, just as important, the stress of producing under those relentless deadlines was *completely eliminated!* Because *I knew* that, no matter how intense the pressure seemed, *I could trust my inner MacGyver to solve the problem.* So the burden was off 'my' shoulders, and all on my inner Mac's—who *always delivered.*

It's hard to describe the *utter sense of relief* I experienced. (And words are my business!) But that gnawing knot of fear in the pit of my stomach every time I had to come up with another story...The anxious and sleepless nights worrying... The mounting tension as I drove to the office knowing I was coming up to bat...**THEY WERE ALL GONE!**

I had unlocked my inner MacGyver and, despite the deadlines, I had never felt *more free or productive in my life.* And my work only got stronger and better with every story and script.

I had broken that insidious cycle of guilt and procrastination that haunts so many of us. Where you can't escape that looming deadline– and keep putting off the work because you're not sure what to do, or fear it won't be good enough—feeling guiltier by the hour for wasting precious time—until, finally cornered by the clock—you desperately throw something together—praying it's not a complete disaster when you have to deliver it.

Gone. GONE. *ALL GONE!*

I could do my best work *and still live my life!* I could fully be there for my wife and my kids and my relatives and my friends because I was no longer in the vice-grip of fear about losing my job. I could truly relax, exhale from the very center of my being and *BREATHE!* Oh yeah, and remember how to laugh out loud.

What's more, relieved of that anxiety and fear, I could see *everything* more clearly: My co-workers and colleagues, how the entertainment business really worked, what I wanted—and *didn't* want—from my career. And that was just the tip of the iceberg.

I began to really **see** and **know** myself in a whole new way. I became more open, more compassionate, and more **aware** of what was going on in my family, my community and the world around me. In a very real sense, the bubble of my personal little universe had dissolved and I was now somehow truly **present**.

In short, I was more *AWAKE* and *ALIVE* than I had ever been.

Now, I have gone to see therapists a number of times in my life when confronted by difficult situations or circumstances. Which sometimes proved helpful. And I would never discourage anyone from seeking professional help if they need it.

But none of those experiences came close to the revelations and directions offered by my inner MacGyver. Because all of that therapy was really a roundabout way to uncover what my inner Mac could tell me, *just by my asking.*

And the personal truth of who I was, what I wanted, and what I needed to do to achieve that was increasingly available to me.

Simply because I could now see past the hamster wheel in my head to all the incredible stuff available to me from my inner MacGyver.

Was my life perfect? Is it now? Of course not. As I said earlier, nobody's life is perfect, and that includes mine. I'm human and so I have problems. Like everyone else.

But now at least I *know there is a way* to confront any challenge or problem in my work or my life. And either resolve it or come to terms with it.

And that, my friends, is a secret worth having.

Chapter 18
What's Next?

As with any new tool offered to this thing we call civilization, the future of the MacGyver Secret will be *written by you*. If, after reading this book, you should try to connect with your inner Mac and add it to your toolbox for solving problems, then your experiences will decide how it is ultimately understood, refined and shared going forward.

So we invite you to *share your experiences with this secret* on MacGyverSecret.com. That way, not only will others benefit from what you learn about it, and Colleen and her counterparts gain valuable info from your insights, but we will be able to adapt and clarify more about how and why this works.

Because, though the secret has some common elements for everyone, *there is no one right way to do it.* The connection between people and their inner MacGyver can be as varied and unique as the number of people who try it. And so I thank you in advance for whatever you are willing to share with the rest of us so the secret can develop and expand in ways none of us have yet imagined.

In turn, we will use the website to continue to share what we learn from any current studies, as well as how it may apply to other professions or endeavors. The conversation doesn't have to end between you and your inner MacGyver. If you choose, it can continue with all of us who use it or are thinking of trying to.

If, as I've suggested, our inner Macs are as vast, powerful and untapped in addressing our individual problems, then imagine what that might mean for solving the mountain of challenges that now confront us all on a global level. One can only dream of the positive outcomes that could come from thousands, or tens of thousands, or perhaps *millions* of us using our inner MacGyvers to offer up truly innovative and workable solutions that we will need to move that mountain.

That's at least one of my dreams. And I welcome you to share that as well.

Acknowledgements

The MacGyver Secret, in all its forms, would not have been possible without the contributions and grace of many people, each of whom deserves more thanks — and praise — than I can adequately express. Suffice it to say then, that they are all bona fide MacGyvers in their own way, and I am supremely fortunate to have had them by my side on this journey.

Jared Krause • Colleen Seifert
Paul Zelizer • Peter Aden • Sasha Allenby
Shawn Patrick • Chris Cade

The MacGyver Secret

CHEAT SHEET

Here are all the KEY STEPS and points of The MacGyver Secret.

- **Write down your question or problem (longhand is best)**
    - Take your time
    - Be as specific as you can
- **Tell your inner MacGyver to work on it for you**
    - Set a time limit
    - Then *let it go*
- **Do some enjoyable or necessary activity**
    - Keep it physical
    - Keep it unimaginative
    - Forget TV, Reading or intense video games
    - Keep talk to a minimum
- **After the time limit, ask your inner MacGyver for the Answers**
    - Return to your written questions
    - Start writing, *anything*
    - Keep writing until you have the answers
- **For Tough or very personal questions**
    - Ask more than once
    - Be patient
- **Try to make connecting with your inner MacGyver a habit**
    - Pick the same time and place for asking
    - Pick the same time and place for receiving the answers

End Notes

CHAPTER 2
PAGE 23: *The Science: What Leads to 'AHA'?*

1. Raichle, M. E., MacLeod, A. M., Snyder, A. Z., Powers, W. J., Gusnard, D. A., Shulman, G. L. (2001). A default mode of brain function. *Proceedings of the National Academy of Sciences of the United States of America, 98*(2):676-682.

2. Kounios, J., Fleck, J. I., Green, D. L., Payne, L., Stevenson, J. L., Bowden, E. M., & Jung-Beeman, M. (2008). The origins of insight in resting-state brain activity. *Neuropsychologia, 46*(1), 281-291.

3. Killingsworth, M. A., & Gilbert, D. T. (2010). A wandering mind is an unhappy mind. *Science, 330*(6006), 932.

4. Kounios, J., & Beeman, M. (2014). The cognitive neuroscience of insight. *Annual Review of Psychology, 65,* 71-93.

4. Christoff, K., Gordon, A. M., Smallwood, J., Smith, R., & Schooler, J. W. (2009). Experience sampling during fMRI reveals default network and executive system contributions to mind wandering. *Proceedings of the National Academy of Sciences of the United States of America, 106*(21), 8719-8724.

5. Salvi, C., Bricolo, E., Franconeri, S. L., Kounios, J., & Meeman, M. (2015). Sudden insight is associated with shutting out visual inputs. *Psychonomic Bulletin & Review, 22*(6), 1814-1819.6

6. Schooler, J. W., Smallwood, J., Christoff, K., Handy, T. C., Reichle, E. D., & Sayette, M. A. (2011). Meta-awareness, perceptual decoupling and the wandering mind. *Trends in Cognitive Sciences, 15*(7), 319-326.

CHAPTER 5
PAGE 34: *The Science: Why Does Writing in Longhand Help?*

1. Mueller, P. A., & Oppenheimer, D. M. (2014). The pen is mightier than the keyboard: Advantages of longhand over laptop note taking. *Psychological Science, 25*(6), 1159-1168.

2. Tulving, E., & Thomson, D. M. (1973). Encoding specificity and retrieval processes in episodic memory. *Psychological Review, 80*(5), 352-373.

3. James, K. H., & Engelhardt, L. (2012). The effects of handwriting experience on functional brain development in pre-literate children. *Trends in Neuroscience and Education, 1*(1), 32–42.

4. Kross, E., Bruehlman-Senecal, E., Park, J., Burson, A., Dougherty, A., Shablack, H., & Bremner, R. (2014). Self-talk as a regulatory mechanism: How you do it matters. *Journal of Personality and Social Psychology, 106*(2), 304-324.

5. Hayes, S. C., Luoma, J. B., Bond, F. W., Masuda, A., & Lillis, J. (2006). Acceptance and commitment therapy: Model, processes and outcomes. *Behaviour Research and Therapy, 44,* 1–25.

CHAPTER 6
PAGE 45: *The Science: Defining the Problem*

1. Csikszentmihalyi, M., & Getzels, J. W. (1971). Discovery-oriented behavior and the originality of creative products: A study with artists. *Journal of Personality Social Psychology, 19,* 47–52.

2. Wallas, G. (1926). *The art of thought.* London: J. Cape

3. Csikszentmihalyi, M., & Getzels, J. W. (1988). Creativity and problem finding in art. In F. Farley & R. Neperud (eds.), *The foundations of aesthetics, art, and art education* (pp. 91-116). New York: Praeger.

4. Dorst, K., & Cross, N. (2001). Creativity in the design process: Co-evolution of problem–solution. *Design Studies, 22*(5), 425–437.

CHAPTER 7
PAGE 59: *The Science: How to Incubate*
1. Wallas, G. (1926). *The art of thought*. London: J. Cape.
2. Dijksterhuis, A., & Strick, M. (2016). A case for thinking without consciousness. *Perspectives on Psychological Science, 11*(1), 117-132.
3. Sio, U. N., & Ormerod, T. C. (2009). Does incubation enhance problem solving? A meta-analytic review. *Psychological Bulletin, 35*(1)94-120.
4. Silveira, J. M. (1971). Incubation: The effect of interruption timing and length on problem solution and quality of problem processing. Unpublished Doctoral Dissertation, University of Oregon, Eugene.
5. Gilhooly, K. J., Georgiou, G. J., Garrison, J., Reston, J. D., & Sirota, M. (2012). Don't wait to incubate: Immediate versus delayed incubation in divergent thinking. *Memory and Cognition, 40,* 966-975.
6. McMahon, K., Sparrow, B., Chatman, L., & Riddle, T. (2011). Driven to distraction: The impact of distracter type on unconscious decision making. *Social Cognition, 29*(6), 683-698.
7. Gilhooly, K. J., Georgiou, G., & Devery, U. (2013). Incubation and creativity: Do something different. *Thinking & Reasoning, 19*(2), 137-149.
8. Baird, B., Smallwood, J., Mrazek, M. D., Kam, J. W., Franklin, M. S., & Schooler, J. W. (2012). Inspired by distraction: Mind wandering facilitates creative incubation. *Psychological Science, 23*(10):1117-22.
9. Strick, M., Dijksterhuis, A., Bos, M. W., Sjoerdsma, A., & van Baaren, R. B (2011). A meta-analysis on unconscious thought effects. *Social Cognition, 29*(6), 738-762.
10. Vigneau, M., Beaucousin, V., Hervé, P.Y., Duffau, H., Crivello, F., Houdé, O., Mazoyer, B., & Tzourio-Mazoyer, N. (2006). Meta-analyzing left hemisphere language areas: Phonology, semantics, and sentence processing. *Neuroimage, 30,* 1414–1432.
11. DeLong, K. A., Urbach, T. P., & Kutas. M. (2005). Probabilistic word pre-activation during language comprehension inferred from electrical brain activity. *Nature Neuroscience, 8,* 1117-1121.
12. Axelrod, V., Bar, M., Rees, G., & Yovel, G. (2015). Neural correlates of subliminal language processing. *Cerebral Cortex, 25*(8), 2160–2169.
13. Ritter. S. M., & Dijksterhuis, A. (2014). Creativity—the unconscious foundations of the incubation period. *Frontiers in Human Neuroscience, 8,* 215.

CHAPTER 8
PAGE 68: *The Science: Thinking Without Awareness*
1. Kihlstrom, J.F. (1987). The cognitive unconscious. *Science, 237*(4821):1445-1452.
2. Bargh, J. A. (2012). The automaticity of everyday life. *Trends in Cognitive Sciences, 16*(12), 593-605.
3. Kahneman, D. (2011). *Thinking, fast and slow*. New York: Farrar Straus Giroux.
4. Sherry, E. C. (1953). Some experiments on the recognition of speech, with one and with two ears. *The Journal of the Acoustical Society of America, 25*(5): 975–979.
5. Wilson, T. D. (2002). *Strangers to ourselves: Discovering the adaptive unconscious*. Cambridge, MA: Belnap Press.
6. Dijksterhuis, A. (2004). Think different: The merits of unconscious thought in preference development and decision making. *Journal of Personality and Social Psychology, 87,* 586–598.
7. Dijksterhuis, A., Bos, M. W., Nordgren, L. F., & Van Baaren, R. B. (2006). On making the right choice: the deliberation-without-attention effect. *Science, 311,* 1005–1007.

8. Dijksterhuis, A., Bos, M. W., van der Leij, A., & van Baaren, R. B. (2009). Predicting soccer matches after unconscious and conscious thought as a function of expertise. *Psychological Science, 20*(11), 1381-1387.
9. Dijksterhuis, A., & Strick, M. (2016). A case for thinking without consciousness. *Perspectives on Psychological Science, 11*(1) 117–132

PAGE 70: Einstein, Edison and Dali: The Genius Trick
1. http://www.wilywalnut.com/Thomas-Edison-Power-Napping.html (Ret.6/6/2016).
1. Dali, S, & Chevalier, H (1992) Dover Fine Art, History of Art Series. *50 Secrets of Magic Craftsmanship.* Courier Corporation, 33-45.

CHAPTER 9
PAGE 78: The Science: Becoming Conscious of the Unconscious
1. Freud, S. (1916-1917). *Introductory lectures on psycho-analysis.* Standard Edition. (15 & 16). London: Hogarth Press.
2. Freud, S. (1900), *The interpretation of dreams, IV and V* (2nd ed.). London: Hogarth Press.
3. Pennebaker, J. W. (1997). Writing about emotional experiences as a therapeutic process. *Psychological Science, 8*(3), 162-266.
4. Brown, K. W., & Ryan, R. M. (2003). The benefits of being present: Mindfulness and its role in psychological well-being. *Journal of Personality and Social Psychology, 84*(4), 822-848.
5. BBC News Magazine (2009, June 10). Five dream discoveries. http://news.bbc.co.uk/go/pr/fr/-/2/hi/uk_news/magazine/8092029.stm (Ret.9/20/2016)
6. Ericsson, K. A., Krampe, R. T., & Tesch-Romer, C. (1993). The role of deliberate practice in the acquisition of expert performance. *Psychological Review, 100*(3), 363–406.

CHAPTER 11
PAGE 95: The Science: How to Build a Habit
1. Gollwitzer, P. M. (1999). Implementation intentions: Strong effects of simple plans. *American Psychologist, 54,* 493–503.
2. Brown, I., Sheeran, P., & Reuber, M. (2009). Enhancing antiepileptic drug adherence: A randomized controlled trial. *Epilepsy & Behavior, 16,* 634–639.
3. Patalano, A. L., & Seifert, C. M. (1997). Opportunistic planning: Being reminded of pending goals. *Cognitive Psychology, 34,* 1-36.
4. Danner, U. N., Aarts, H., & de Vries, N. K. (2007). Habit formation and multiple means to goal attainment: Repeated retrieval of target means causes inhibited access to competitors. *Personality and Social Psychology Bulletin, 33,* 1367–1379.
5. Lally, P., Van Jaarsveld, C. H., Potts, H. W., & Wardle, J. (2010). How are habits formed: Modelling habit formation in the real world. *European Journal of Social Psychology, 40,* 998–1009.

CHAPTER 12
PAGE 100: The Science: Using Sleep to Incubate
1. Wagner, U., Gais, S., Haider, H., Verleger, R., & Born, J. (2004). Sleep inspires insight. *Nature, 427,* 352–355.
2. Sio, U. N., Monaghan, P., & Ormerod, T. (2012). Sleep on it, but only if it is difficult: Effects of sleep on problem solving. *Memory & Cognition, 41,* 159–166.
3. Manni, R. (2005). Rapid eye movement sleep, non-rapid eye movement sleep, dreams and hallucinations. *Current Psychiatry Reports, 7,* 196–200.

4 Cai, D. J., Mednick, S. A., Harrison, E. M., Kanady, J. C., & Mednick, S. C. (2009). REM, not incubation, improves creativity by priming associative networks. *Proceedings of the National Academy of Sciences of the United States of America, 106*(25), 10130–10134.

PAGE 101: *Jack Nicklaus, Elias Howe & Niels Bohr: Finding Solutions in a Dream*

1 Happy Dream for Nicklaus: Jack's Back in Form, *San Francisco Associated Press*, June 27, 1964, 37
1 Modern Mysticism: Jung, Zen and the Still Good Hand of God https://books.google.com/books?id=4d4a182uvDkC&pg=PT44&lpg=PT44&dq=San+Francisco+Chronicle+June+27+1964+Jack&source=bl&ots=MQsWhU3b44&sig=FaJPjDogq_IuLJpEC8O47m-zQWs&hl=en&sa=X&ved=0ahUKEwit8CSoeTMAhVkxYMKHVAJAw0Q6AEIKjAC#v=onepage&q=San%20Francisco%20Chronicle%20June%2027%201964%20Jack&f=false, (6/9/2016).
2 http://www.dreaminterpretation-dictionary.com/famous-dreams-elias-howe.html, (Ret. 6/9/2016).
3 Ottaviani, J, (2009) *Suspended In Language: Niels Bohr Life, Discoveries, And The Century He Shaped.* (2nd ed.). G.T. Labs.
3 http://www.world-of-lucid-dreaming.com/10-dreams-that-changed-the-course-of-human-history.html, (Ret. 6/9/2016).

CHAPTER 15
PAGE 117: *The Science: Is Group Brainstorming Better?*

1 Diehl, M., & Stroebe, W. (1987). Productivity loss in brainstorming groups: Toward the solution of a riddle. *Journal of Personality and Social Psychology, 53*(3), 497-509.
2 Mullen, B., Johnson, C., & Salas, E. (1991). Productivity loss in brainstorming groups: A meta-analytic integration. *Basic and Applied Social Psychology, 12*(1), 3-23.
3 Paulus, P. B., Dzindolet, M. T., Poletes, G., & Camacho, L. M. (1993). Perception of performance in group brainstorming: The illusion of group productivity. *Personality & Social Psychology Bulletin, 19*(1), 78-89.

CHAPTER 16
PAGE 123: *The Science: The Prepared Mind*

1 Ritter, S. M., & Dijksterhuis, A. (2014). Creativity—the unconscious foundations of the incubation period. *Frontiers in Human Neuroscience, 8*, 215. http://dx.doi.org/10.3389/fnhum.2014.00215
2 Posner, M. (1973). *Cognition: An introduction.* Glenview, IL: Scott, Foresman (p. 148).
3 Seifert, C. M., Meyer, D. E., Davidson, N., Patalano, A. L., & Yaniv, I. (1995). Demystification of cognitive insight: Opportunistic assimilation and the prepared-mind hypothesis. In R. J. Sternberg & J. E. Davidson (eds.), *The nature of insight* (pp. 65-124). Cambridge, MA: MIT Press.
4 Patalano, A. L., & Seifert, C. M. (1997). Opportunistic planning: Being reminded of pending goals. *Cognitive Psychology, 34*, 1-36

CHAPTER 17
PAGE 127: *Life With Your Inner MacGyver*

1 Jarosz, A., Colflesh, G., & Wiley, J. (2012). Uncorking the muse: Alcohol intoxication facilitates creative problem solving. *Consciousness and Cognition, 21*(1), 487-493

Bibliography

Axelrod, V., Bar, M., Rees, G., & Yovel, G. (2015). Neural correlates of subliminal language processing. *Cerebral Cortex, 25*(8), 2160–2169.

Baird, B., Smallwood, J., Mrazek, M. D., Kam, J. W., Franklin, M. S., & Schooler, J. W. (2012). Inspired by distraction: Mind wandering facilitates creative incubation. *Psychological Science, 23*(10):1117-22.

Bargh, J. A. (2012). The automaticity of everyday life. *Trends in Cognitive Sciences, 16*(12), 593-605.

BBC News Magazine (2009, June 10). Five dream discoveries. http://news.bbc.co.uk/go/pr/fr/-/2/hi/uk_news/magazine/8092029.stm (Ret. 9/20/2016).

Brown, K. W., & Ryan, R. M. (2003). The benefits of being present: Mindfulness and its role in psychological well-being. *Journal of Personality and Social Psychology, 84*(4), 822-848.

Brown, I., Sheeran, P., & Reuber, M. (2009). Enhancing antiepileptic drug adherence: A randomized controlled trial. *Epilepsy & Behavior, 16,* 634–639.

Cai, D. J., Mednick, S. A., Harrison, E. M., Kanady, J. C., & Mednick, S. C. (2009). REM, not incubation, improves creativity by priming associative networks. *Proceedings of the National Academy of Sciences of the United States of America, 106*(25), 10130–10134.

Christoff, K., Gordon, A. M., Smallwood, J., Smith, R., & Schooler, J. W. (2009). Experience sampling during fMRI reveals default network and executive system contributions to mind wandering. *Proceedings of the National Academy of Sciences of the United States of America, 106*(21), 8719-8724.

Csikszentmihalyi, M., & Getzels, J. W. (1971). Discovery-oriented behavior and the originality of creative products: A study with artists. *Journal of Personality Social Psychology, 19,* 47–52.

Csikszentmihalyi, M., & Getzels, J. W. (1988). Creativity and problem finding in art. In F. Farley & R. Neperud (eds.), *The foundations of aesthetics, art, and art education* (pp. 91-116). New York: Praeger.

Dali, S, & Chevalier, H (1992) Dover Fine Art, History of Art Series. *50 Secrets of Magic Craftsmanship.* Courier Corporation, 33-45.

Danner, U. N., Aarts, H., & de Vries, N. K. (2007). Habit formation and multiple means to goal attainment: Repeated retrieval of target means causes inhibited access to competitors. *Personality and Social Psychology Bulletin, 33,* 1367–1379.

DeLong, K. A., Urbach, T. P., & Kutas. M. (2005). Probabilistic word pre-activation during language comprehension inferred from electrical brain activity. *Nature Neuroscience, 8,* 1117-1121.

Diehl, M., & Stroebe, W. (1987). Productivity loss in brainstorming groups: Toward the solution of a riddle. *Journal of Personality and Social Psychology, 53*(3), 497-509.

Dijksterhuis, A. (2004). Think different: The merits of unconscious thought in preference development and decision making. *Journal of Personality and Social Psychology, 87,* 586–598.

Dijksterhuis, A., Bos, M. W., Nordgren, L. F., & Van Baaren, R. B. (2006). On making the right choice: the deliberation-without-attention effect. *Science, 311,* 1005–1007.

Dijksterhuis, A., Bos, M. W., van der Leij, A., van Baaren, R. B. (2009). Predicting soccer matches after unconscious and conscious thought as a function of expertise. *Psychological Science, 20*(11), 1381-1387.

Dijksterhuis, A., & Strick, M. (2016). A case for thinking without consciousness. *Perspectives on Psychological Science, 11*(1), 117-132.

Dorst, K., & Cross, N. (2001). Creativity in the design process: Co-evolution of problem–solution. *Design Studies, 22*(5), 425–437.

Ericsson, K. A., Krampe, R. T., & Tesch-Romer, C. (1993). The role of deliberate practice in the acquisition of expert.

Freud, S. (1900), *The interpretation of dreams, IV and V* (2nd ed.). London: Hogarth Press.

Freud, S. (1916-1917). *Introductory lectures on psycho-analysis.* Standard Edition. (15 & 16). London: Hogarth Press.

Gilhooly, K. J., Georgiou, G., & Devery, U. (2013). Incubation and creativity: Do something different. *Thinking & Reasoning, 19*(2), 137-149.

Gilhooly, K. J., Georgiou, G. J., Garrison, J., Reston, J. D., & Sirota, M. (2012). Don't wait to incubate: Immediate versus delayed incubation in divergent thinking. *Memory and Cognition, 40,* 966-975.

Gollwitzer, P. M. (1999). Implementation intentions: Strong effects of simple plans. *American Psychologist, 54,* 493–503.

Hayes, S. C., Luoma, J. B., Bond, F. W., Masuda, A., & Lillis, J. (2006). Acceptance and commitment therapy: Model, processes and outcomes. *Behaviour Research and Therapy, 44,* 1–25.

James, K. H., & Engelhardt, L. (2012). The effects of handwriting experience on functional brain development in pre-literate children. *Trends in Neuroscience and Education, 1*(1), 32–42.

Jarosz, A., Colflesh, G., & Wiley, J. (2012). Uncorking the muse: Alcohol intoxication facilitates creative problem solving. *Consciousness and Cognition, 21*(1), 487-493

Kahneman, D. (2011). *Thinking, fast and slow.* New York: Farrar Straus Giroux.

Kihlstrom, J .F. (1987). The cognitive unconscious. *Science, 237*(4821):1445-1452.

Killingsworth, M. A., & Gilbert, D. T. (2010). A wandering mind is an unhappy mind. *Science, 330*(6006), 932.

Kounios, J., & Beeman, M. (2014). The cognitive neuroscience of insight. *Annual Review of Psychology, 65,* 71-93.

Kounios, J., Fleck, J. I., Green, D. L., Payne, L., Stevenson, J. L., Bowden, E. M., & Jung-Beeman, M. (2008). The origins of insight in resting-state brain activity. *Neuropsychologia, 46*(1), 281-291.

Kross, E., Bruehlman-Senecal, E., Park., J., Burson, A., Dougherty, A., Shablack, H., & Bremner, R. (2014). Self-talk as a regulatory mechanism: How you do it matters. *Journal of Personality and Social Psychology, 106*(2), 304-324.

Lally, P., Van Jaarsveld, C. H., Potts, H. W., & Wardle, J. (2010). How are habits formed: Modelling habit formation in the real world. *European Journal of Social Psychology, 40,* 998–1009.

McMahon, K., Sparrow, B., Chatman, L., & Riddle, T. (2011). Driven to distraction: The impact of distracter type on unconscious decision making. *Social Cognition, 29*(6), 683-698.

Manni, R. (2005). Rapid eye movement sleep, non-rapid eye movement sleep, dreams and hallucintions. *Current Psychiatry Reports, 7,* 196–200.

Mullen, B., Johnson, C., & Salas, E. (1991). Productivity loss in brainstorming groups: A meta-analytic integration. *Basic and Applied Social Psychology, 12*(1), 3-23.

Mueller, P. A., & Oppenheimer, D. M. (2014). The pen is mightier than the keyboard: Advantages of longhand over laptop note taking. *Psychological Science, 25*(6), 1159-1168.

Ottaviani, J, (2009) *Suspended In Language: Niels Bohr Life, Discoveries, And The Century He Shaped.* (2nd ed.). G.T. Labs.

Patalano, A. L., & Seifert, C. M. (1997). Opportunistic planning: Being reminded of pending goals. *Cognitive Psychology, 34,* 1-26.

Paulus, P. B., Dzindolet, M. T., Poletes, G., & Camacho, L. M. (1993). Perception of performance in group brainstorming: The illusion of group productivity. *Personality & Social Psychology Bulletin, 19*(1), 78-89.

Pennebaker, J. W. (1997). Writing about emotional experiences as a therapeutic process. *Psychological Science, 8*(3), 162-266.

Posner, M. (1973). *Cognition: An introduction.* Glenview, IL: Scott, Foresman (p. 148).

Raichle, M. E., MacLeod, A. M., Snyder, A. Z., Powers, W. J., Gusnard, D. A., Shulman, G. L. (2001). A default mode of brain function. *Proceedings of the National Academy of Sciences of the United States of America, 98*(2):676-682.

Ritter. S. M., & Dijksterhuis, A. (2014). Creativity—the unconscious foundations of the incubation period. *Frontiers in Human Neuroscience, 8,* 215. http://dx.doi.org/10.3389/fnhum.2014.00215

Salvi, C., Bricolo, E., Franconeri, S. L., Kounios, J., & Meeman, M. (2015). Sudden insight is associated with shutting out visual inputs. *Psychonomic Bulletin & Review, 22*(6), 1814-1819.6

Schooler, J. W., Smallwood, J., Christoff, K., Handy, T. C., Reichle, E. D., & Sayette, M. A. (2011). Meta-awareness, perceptual decoupling and the wandering mind. *Trends in Cognitive Sciences, 15*(7), 319-326.

Seifert, C. M., Meyer, D. E., Davidson, N., Patalano, A. L., & Yaniv, I. (1995). Demystification of cognitive insight: Opportunistic assimilation and the prepared-mind hypothesis. In R. J. Sternberg & J. E. Davidson (eds.), *The nature of insight* (pp. 65-124). Cambridge, MA: MIT Press.

Sherry, E. C. (1953). Some experiments on the recognition of speech, with one and with two ears. *The Journal of the Acoustical Society of America, 25*(5): 975–979.

Silveira, J. M. (1971). Incubation: The effect of interruption timing and length on problem solution and quality of problem processing. Unpublished Doctoral Dissertation, University of Oregon, Eugene.

Sio, U. N., Monaghan, P., & Ormerod, T. (2012). Sleep on it, but only if it is difficult: Effects of sleep on problem solving. *Memory & Cognition, 41,* 159–166.

Sio, U. N., & Ormerod, T. C. (2009). Does incubation enhance problem solving? A meta-analytic review. *Psychological Bulletin, 35*(1):94-120.

Strick, M., Dijksterhuis, A., Bos, M. W., Sjoerdsma, A., & van Baaren, R. B (2011). A meta-analysis on unconscious thought effects. *Social Cognition, 29*(6), 738-762.

Tulving, E., & Thomson, D. M. (1973). Encoding specificity and retrieval processes in episodic memory. *Psychological Review, 80*(5), 352-373.

Vigneau, M., Beaucousin, V., Hervé, P.Y., Duffau, H., Crivello, F., Houdé, O., Mazoyer, B., & Tzourio-Mazoyer, N. (2006). Meta-analyzing left hemisphere language areas: Phonology, semantics, and sentence processing. *Neuroimage, 30,* 1414–1432.

Wagner, U., Gais, S., Haider, H., Verleger, R., & Born, J. (2004). Sleep inspires insight. *Nature, 427,* 352–355.

Wallas, G. (1926). *The art of thought.* London: J. Cape.

Wilson, T. D. (2002). *Strangers to ourselves: Discovering the adaptive unconscious.* Cambridge, MA: Belnap Press.

ADDITONAL SOURCES

Happy Dream for Nicklaus: Jack's Back in Form, *San Francisco Associated Press,* June 27, 1964, 37

Modern Mysticism: Jung, Zen and the Still Good Hand of God
https://books.google.com/books?id=4d4a182uvDkC&pg=PT44&lpg=PT44&dq=San+Francisco+C hronicle+June+27+1964+Jack&source=bl&ots=MQsWhU3b44&sig=FaJPjDogq_IuLJpEC8O47m-zQWs&hl=en&sa=X&ved=0ahUKEwit8CSoeTMAhVkxYMKHVAJAw0Q6AEIKjAC#v=onepage&q=San %20Francisco%20Chronicle%20June%2027%201964%20Jack&f=false, (Ret. 6/9/2106).

http://www.dreaminterpretation-dictionary.com/famous-dreams-elias-howe.html, (Ret.6/9/2016).

http://www.wilywalnut.com/Thomas-Edison-Power-Napping.html, (Ret. 6/6/2016).

http://www.world-of-lucid-dreaming.com/10-dreams-that-changed-the-course-of-human-history.html, (Ret. 6/9/2016).

Index
Science & Stories

SCIENCE

What Leads to 'AHA'? 23
Why Does Writing in Longhand Help? 34
Defining the Problem 45
How to Incubate 59
Thinking Without Awareness 68
Becoming Conscious of the Unconscious 78
How to Build a Habit 95
Using Sleep to Incubate 100
Is Group Brainstorming Better? 117
The Prepared Mind 123

AHA

How I Discovered The MacGyver Secret 11
How I Put My Inner MacGyver to Work 15
Jared's Story 36
Colleen's Story, *Science Advisor* 46
Harley's Story 62
Einstein, Edison and Dali: *The Genius Trick* 70
Michelle's Story 83
Sally's Story 96
Jack Nicklaus, Elias Howe & Niels Bohr: *Finding Solutions in a Dream.* 101
Allen's Story 107

About the Authors

Lee D. Zlotoff

Lee Zlotoff is an award-winning writer, producer and director of film and television with over a hundred primetime credits to his name. Among his more notable creations is the iconic MacGyver TV series that has continued to run around the world for over 30 years since its debut in 1985. MacGyver is now returning in a new series on CBS and on an upcoming feature film from Lionsgate Studios. Lee is also a recipient of the coveted Audience Award from the Sundance Film Festival for his film *'The Spitfire Grill'* which has gone on to become a highly successful musical.

Lee is a graduate of *St. John's College,* with campuses in Annapolis, MD and Santa Fe. NM, where he served on the board of directors for over a decade and is now a member of the President's Council.

As the father of four grown children, and now the grandfather of four, Lee is committed to supporting those who strive to improve global futures and outcomes. This is why he—and his children—have formed *The MacGyver Foundation,* which receives a portion of the proceeds from every MacGyver project.

Lee currently lives in Santa Fe, New Mexico with his partner, the accomplished artist and painter, Dayna Matlin.

Colleen M. Seifert, Ph.D.

Dr. Colleen M. Seifert is the SCIENCE ADVISOR for *The MacGyver Secret.* As a cognitive scientist, she offers expertise informing the development of the book and workshops.

Colleen is an Arthur F. Thurnau Professor at the *University of Michigan, Ann Arbor.* She received her Ph.D. in Psychology from *Yale University.* After a postdoctoral fellowship, Colleen moved to the *University of Michigan* in Ann Arbor, was promoted to the rank of full professor in 2001.

Colleen has published over fifty scholarly articles addressing the intersections of thinking, memory, and creativity. Her research on problem solving and the creative process has led to methods to enhance creativity both in design and education. She is a past President and Executive Officer of the *Cognitive Science Society,* and a Fellow in the *American Psychological Society.*

Colleen lives with her husband, Zeke Montalvo, and son, Victor, in Ann Arbor, Michigan.

Full Online Training Course

For those who would like to experience the most powerful and enduring results from

The MacGyver Secret

Using video based instruction and experiential exercises, Lee Zlotoff will guide you through this **12+ week course,** going even deeper into the topics of the book and the science.

The net result is to weave the Secret seamlessly and indelibly into your work and life. You will learn how to easily and successfully resolve any question or problem, whether technical, creative, professional or personal.

Speaking and Workshops

Lee Zlotoff is an accomplished speaker and workshop presenter who engages business, academic, professional and community audiences. He has taught *The MacGyver Secret* at major universities, research institutes, and corporations, including: *Harvard* and *Stanford*, the *Jet Propulsion Laboratory* in Pasadena, CA, *Zappos Inc.* and *The Naspers Media Company,* just to name a few.

www.MacGyverSecret.com

CPSIA information can be obtained
at www.ICGtesting.com
Printed in the USA
LVOW13s1441101116
512465LV00009B/531/P